Classification and Evolution in Biology, Linguistics and the History of Science
Edited by Heiner Fangerau, Hans Geisler, Thorsten Halling
and William Martin

KULTUR ANAMNESEN

Schriften zur Geschichte und Philosophie der Medizin und der Naturwissenschaften

Herausgegeben von Heiner Fangerau, Renate Breuninger und Igor Polianski
in Verbindung mit dem Institut für Geschichte, Theorie und Ethik der Medizin, dem Humboldt-
Studienzentrum für Philosophie und Geisteswissenschaften und dem Zentrum Medizin
und Gesellschaft der Universität Ulm

Band 5

Classification and Evolution in Biology, Linguistics and the History of Science

Concepts – Methods – Visualization

EDITED BY HEINER FANGERAU, HANS GEISLER,
THORSTEN HALLING AND WILLIAM MARTIN

Franz Steiner Verlag

Based on a project funded by the:

Federal Ministry
of Education
and Research

Umschlagabbildungen:
Reihenlogo: Walter Draesner, „Der Tod und der Anatom", Graphiksammlung „Mensch und Tod"
der Heinrich-Heine-Universität Düsseldorf
Abbildung: "Network" by Arno Görgen

Bibliografische Information der Deutschen Nationalbibliothek:
Die Deutsche Nationalbibliothek verzeichnet diese Publikation in der Deutschen
Nationalbibliografie; detaillierte bibliografische Daten sind im Internet über
<http://dnb.d-nb.de> abrufbar.

TABLE OF CONTENTS

FOREWORD

Shortly after Charles Darwin published his seminal work *On the Origin of Species* in 1859, the concept of "evolution" entered nineteenth century thinking, and soon became a general metaphor to describe developmental processes in many scientific disciplines. One of the first scientists to adapt to Darwinian ideas was August Schleicher, who, in an open letter to Ernst Haeckel (1863), pointed out striking similarities between linguistic and biologic descent. He was also the first to present family trees as evolutionary trees, exemplified by postulating a common ancestor of all Indo-European languages. In 1871, Darwin incorporated these proposals in his book entitled *The Descent of Man and Selection in Relation to Sex*, in which he placed strong emphasis on the importance of natural selection in linguistic evolution.[1] The on-going debates about evolution in biology and comparative philology had major cross-disciplinary impacts on theory building, both in natural and cultural sciences, and finally gave rise to "Universal Darwinism" (Dawkins 1983). Within the framework of a comprehensive "generalised theory of evolution," the Darwinian principles of reproduction, variation and selection have gradually become detached from their biological substrate, being construed as abstract properties of dynamic systems (for summaries see Gontier et al. 2006; Schurz 2011; Mesoudi 2011; Brinkworth et al. 2012; Ruse 2012; Sydow 2012).

Recently, there has been an increase in the number of critics of universal or "generalised Darwinism", who view it "as an overarching research strategy" (Levit et al. 2011). Specifically, critics have questioned the explanatory power of this approach, which is based on the assumption of a fundamental homology between evolution in nature and the evolution of any kind of culture.

While Darwinism has undergone many changes, and shown up in many facets, there remains an outstanding common feature in its history spanning more than 150 years; since the very beginning, branching trees have been the dominant scheme for representing evolutionary processes. In the analogy with kinship relations in a family tree, this scheme exclusively models evolution as vertical inheritance. However, the scheme does not cover lateral transfer, that is, the mixing or hybridizing species or languages. To describe this latter phenomenon, a reasonable approach seems to be the use of the network metaphor.

Different from powerful bifurcating tree graphs, the use of network graphs to represent the development of species and languages has only recently received increasing interest in the fields of science and humanity; even if networks may be traced back to the eighteenth century in both linguistics and biology. Today, models of reticulation are widely used in a variety of scientific fields on a formalized basis.

---

1   "The formation of different languages and of distinct species, and the proofs that both have been developed through a gradual process, are curiously parallel. ... The survival or preservation of certain favoured words in the struggle for existence is natural selection" (Darwin 1882: 90).

In biology, research on prokaryote evolution indicates that lateral gene transfer is a major feature in the evolution of bacteria. In the field of linguistics, the mutual lexical and morphosyntactic borrowing between languages, as well as the wave-like distribution of innovations, seems to be much more central for language evolution, as the family tree model is likely to concede. In the humanities, networks are employed as an alternative to established phylogenetic models, to express the hybridisation of cultural phenomena, concepts or the social structure of science.

However, an interdisciplinary display of network analyses for evolutionary processes remains lacking. It is this gap we intend to fill with our book. The book is directed towards a wide readership, including biologists, who are interested in the methodological and theoretical reflections of evolution, linguists, who work on the development of languages, and historians of science, who examine the evolution of ideas. This book is based on an interdisciplinary conference and an interdisciplinary research project that were funded by the German Ministry of Education, and which focused on examining the concepts of evolutionary processes in different disciplines from a general perspective. However, these concepts were not regarded as completely homogeneous, but comparable according to similar relationship patterns. Therefore, this volume includes approaches studying the evolutionary dynamics of science, languages and genomes, all of which were based on methods incorporating network approaches.

We wish to thank all contributors, and hope to foster research in the direction of evolution that is understood as a network process in different fields of research.

The Editors, May 2013

REFERENCES

Brinkworth, M. & Weinert, F. (2012): Evolution 2.0. Implications of Darwinism in Philosophy and the Social and Natural Sciences (Berlin, Heidelberg: Springer).

Darwin, C. (1859): On the Origin of species by means of natural selection: or the preservation of favored races in the struggle for life (London: John Murray)

Darwin, C. (1882 [1871]): The Descent of Man and Selection in Relation to Sex, (2nd ed), (London: John Murray).

Dawkins, R. (1983): 'Universal Darwinism' in D.S Bendall (ed), Evolution from Molecules to Man (Cambridge, UK: Cambridge University Press): 403–425

Gontier, N., Van Bendegem, J.P. & Aerts, D. (eds) (2006): Evolutionary Epistemology, Language and Culture: a Non-Adaptationist Systems Theoretical Approach (Dordrecht: Springer).

Levit, G.S., Hossfeld, U. & Witt, U. (2011): 'Can Darwinism be "Generalized" and of what use would this be?', Journal of Evolutionary Economnics 21:545–562.

Mesoudi, A. (2011): Cultural Evolution: How Darwinian Theory Can Explain Human Culture and Synthesize the Social Sciences (Chicago: Chicago University Press).

Ruse, M. (ed) (2012), The Cambridge encyclopedia of Darwin and evolutionary thought (Cambridge: Cambridge Univ. Press), 2012.

Schurz, G. (2011): Evolution in Natur und Kultur. Eine Einführung in die verallgemeinerte Evolutionstheorie (Heidelberg: Spektrum).

Sydow, M.v. (2012) From Darwinian metaphysics towards understanding the evolution of evolutionary mechanisms. A historical and philosophical analysis of gene-*Darwinism* and universal *Darwinism* (Göttingen: Universitätsverlag Göttingen).

1. NETWORKS AND EVOLUTION
IN THE HISTORY OF SCIENCE

EVOLUTION OF KNOWLEDGE FROM A NETWORK PERSPECTIVE: RECOGNITION AS A SELECTIVE FACTOR IN THE HISTORY OF SCIENCE

Heiner Fangerau

"Darwinian" approaches to describe the development of knowledge gained wide public reception in the 1970s and 1980s, when several books about connections between biological evolution and the evolution of concepts in science were published and when corresponding ideas of leading authorities in biology and philosophy, like Konrad Lorenz or Karl Popper, were popularized.[1] In this context Donald T. Campbell (1974) coined the term "evolutionary epistemology" in an essay about Popper's theories of conceptual change to refer to this interdisciplinary endeavour to find generalising descriptions of knowledge development. He interpreted Popper's ideas in light of metaphors borrowed from evolutionary biology and argued convincingly that the development of scientific knowledge was the result of variation, trial and error, transmission, selection, and adaptation (Campbell 1974).

Of course, the basic conceptual link between epistemological considerations and the theory of evolution is much older and can be dated back at least to the 19th century (Richards 1987: 575), but Campbell's introduction of this term commenced a lasting debate about the strength and validity of analogising knowledge development and biology. It soon became clear that the meaning of "evolutionary epistemology" needed clarification, especially because Lorenz and Popper seemed to have addressed different spheres of interest when they replied to Campbell's ideas (Vollmer 1987). Lorenz addressed the "evolution of cognitive systems in general and of our cognitive abilities in particular" (Vollmer 1987: 203), whereas Popper discussed the evolution of scientific knowledge. He was interested in the philosophical and historical aspects of the development and fate of scientific knowledge, rather than in the biological foundations of the brain's cognitive functions. Following this direction, which has been thoroughly discussed by authors such as Stephen Toulmin (1972), Robert Richards (1987), and David Hull (2001 [1988]), the aim of

---

1 See for example Oeser (1988); Plotkin (1982); Popper (1979); Radnitzky and Bartley (1987); Reitmeyer and Marx (2010); Richards (1987); Riedl and Kaspar (1980); Vollmer (1975); Wuketits (1983). A dialogue touching the issue held by Lorenz and Popper in Altenburg in 1983 was published as a pocket book and sold in 6 editions with 36.000 copies until 1994 (Popper et al. 1985: 30–31). A review of the German discourse was published in the weekly newspaper "Die Zeit" in 1980 calling evolutionary epistemology the Copernican turn of our times (Zimmer 06.06.1980).

this paper is to add a network perspective to the evolutionary interpretation of the history of science.[2]

I will argue that knowledge development can be reconstructed and displayed as a networking process. In this approach networks are characterized by nodes representing entities like ideas or people, links symbolising horizontal and vertical relations between nodes and the absence of a clear staring point. With this approach, I take a path paved by authors such as Stephen Toulmin, Mary Hesse and Bruno Latour, who described different network perspectives on science and the scientific system around the time that evolutionary epistemology entered the scientific discourse.[3] Toulmin for example addressed "the rational enterprise of a natural science [...] as a changing population of scientists, linked together in more or less formally organized institutions" (Toulmin 1972: 262) and Hesse developed a network model which "interprets scientific theory in terms of a network of concepts related by laws, in which only pragmatic and relative distinctions can be made between the 'observable' and the 'theoretical'" (Hesse 1974: 4). On cognitive and social levels, network analyses can be used to reconstruct the evolution of scientific ideas as elements of scientific concepts. The actors responsible for the processes of selection and transfer constitute the organisational structure of such a network, in which knowledge is produced, retained, and transmitted through lateral and horizontal transfer. In analogy to biology, Hull called this formation the demic structure of science (Grantham 2000; Hull 2001 [1988]). Thus, I follow the suggestion that the evolution of knowledge as a social product can be better described and examined using the "network" metaphor than with the classical tree-like model of vertical transgenerational transfer borrowed from classical biological concepts of evolution.[4] In this approach, "selection" may be viewed as the evolutionary element – certain replicators (ideas, approaches, theories) that fit the environment (the rationality and motivation of interactors) are chosen from a variety of possibilities – and the network can be viewed as a representation format for the reconstruction of lateral transfer in the histories of medicine and biology.

I will also argue – based on a historical example – that the processes of selection and transfer, which occur in current biological and medical research, are partly the result of researchers' personal motivations to gain recognition for their scientific work within a network of scientists.[5] Additionally, the issue of which researcher other scientists believe in a controversial situation depends to a certain extent on whom they trust for any reason. Thus, I suggest that "recognition" and related "self-

---

2   In his proposal of "A neo-Darwinian model of science", Knudsen (2003) provided an excellent short overview of the differences and commonalities among these authors.
     For an overview of Evolutionary Epistemology see also the essay collection by Radnitzky and Bartley (1987), which also includes essays by Popper, Campbell and Vollmer.

3   See, for example, Hesse (1974), Latour (2005). For a precise and concise overview, see Dear (2012: 46–50).

4   Molecular evolutionists have argued that the network approach is superior for the description of biological evolution because it enables, for example, the characterisation of processes of lateral gene transfer. For an overview, see Martin (2011).

5   Some of my thoughts on recognition in science presented here and in the following have been recently published also in German (Krischel, Halling, Fangerau 2012).

constitution" are driving forces in the evolution of knowledge in networks. These factors can be seen as crucial elements of selection and transfer when these processes are understood as being organised within a social structure of science.

After a very brief overview of the key features and limits of an evolutionary epistemology I propose the use of a network approach to map the evolution of ideas: connections among scientists are described as representations of the replication of ideas, the diachronic perspective on these connections helps to characterise knowledge evolution as a networking process. Following these theoretical considerations based on the existing literature, I use the example of the physiologist Jacques Loeb's views on citation around 1900 to describe how scientists' desire for recognition drives the selection and transfer of scientific ideas. By connecting this empirical example to the previous considerations, I finally take up the idea that recognition and self-constitution are important driving forces in networking processes in science and argue that they foster the evolution of knowledge.

EVOLUTIONARY EPISTEMOLOGY

Some factors support the belief that the historical aspects of the development of ideas can indeed be understood in evolutionary terms. Most systems evolve (Vollmer 1987: 212), and several metaphors from biology can be transferred readily to the description of knowledge development. Above all, scientists' selection of ideas, theories, and/or concepts and their transfer to others, sometimes transgenerationally, is a central element of the system of science. Variations and transformations of ideas occur and the recombination of ideas generates new concepts. Hypotheses can be seen as replicators during this process (Popper 1979), and the scientists involved as interactors (Hull 2001 [1988]). In concordance with the topos of the "survival of the fittest", realists or empiricists have suggested that theories fitting reality best or according with empirical observations "survive" selection and are transferred (Collin 2003). Similarly, Niklas Luhmann (1998: 546–56) viewed the evolutionary selection of ideas as resting on criteria of plausibility or self-evidence, but, in contrast to the realists, he pointed out the historical contingency of what is considered to be plausible.

Nevertheless, many authors have cautioned that efforts to equate organic evolution with the development of science as a system in general and the evolution of scientific knowledge in particular may be too hasty. Above all, selection processes in biology and the system of science differ. For example, Vollmer (1987: 214) warned against equating fitness, the evolutionary criterion for success, with the "truth" of scientific knowledge because "fitness may be provided by quite limited or even deceptive cognitive means". Additionally, human influence on science is much greater than on natural (not breeding) biological selection processes, and scientists' motivations must be taken into account.[6] Finally, the evolution of ideas seems to be more goal oriented (explaining phenomena on the basis of rationality

---

6   This argument can be traced back to the psychological theory of Adam Smith (Loasby 2002).

and defined methodology) than undirected biological evolution, which is comparable to trial and error (Sterelny 1994). As Paul Thagard (1980: 193) pointed out when referring to the evolution of knowledge,

> "Variation is not blind [...] it is not wholly [...] determined by context either. There is a subjective, psychological element in discovery along with an aim-oriented, methodological element. Hence we are not in a position to borrow a model for the growth of knowledge from Lamarck, Hegel, or Darwin."

He stated that publication and pedagogy, rather than a process similar to biological inheritance, were the forces driving the transmission and preservation of knowledge (Thagard 1980: 192). He urged the development of a model that included

> "1. the intentional, abductive activity of scientists in initially arriving at new theories and concepts; 2. the selection of theories according to criteria which reflect general aims; 3. the achievement of progress by sustained application of criteria; and 4. the rapid transmission of selected theories in highly organized scientific communities" (Thagard 1980: 193).

Following this stream of thought, it might be argued from a historical standpoint for an evolutionary epistemology that focuses on "selection" and "transfer" as crucial elements in the development of knowledge.[7] That said, I do not discard analogies between biological evolution and the evolution of science, but suggest retaining them on the broad level of the metaphorical explanation of mechanisms that work in systems. I am aware that I take an explicitly "externalist" perspective with this suggestion, as I do not examine theories or hypotheses alone, although they are substantial elements of scientific endeavours. Rather, the relational and social aspects of the methodologically guided production of knowledge are focussed here. With this emphasis, it is not intended to argue against realists' claims that scientific knowledge has a counterpart in the real world or that scientists are working to find the "truth" (Churchland and Hooker 1985). Rather, I propose concentrationg on the networks of the producers of this knowledge and the social mechanisms of selecting and transferring special representations of viewing the phenomena of the world.

MAPPING THE EVOLUTION OF CONCEPTS: A NETWORK APPROACH

Above all, a network is a graphical representation of relationships, "a collection of points joined together in pairs by lines" (Newman 2010: 1). It consists of nodes (or vertices) representing elements that are linked and links (or edges) representing different forms of connection. The whole system of nodes and links is called a graph. Links between nodes can have different strengths and nodes can be closely related *via* other nodes without being linked directly. Thus, a network is an overarching description of connected elements, or, as Easley and Kleinberg (2010: 1) described it, "a pattern of interconnections among a set of things". This description captures the semantic connotation of the term "network" more than the pure graphical de-

---

7   Selection and transfer, or transmission, are linked, as suggested by Knudsen's (2003: 103) definition of the "[...] selection of explicit scientific knowledge as the gradual and slow change in the distribution of scientific ideas caused by their differential social transmission".

scription. Network analyses have a long tradition in sociology, where they have been used to describe the structures of social relations and the regularities between certain relational structures and various kinds of social interaction or effect, such as the exertion of power or economic success (Freeman 2004). However, network models can be adapted to "data that do not reflect concrete social relations but rather relations among concepts or discursive elements" (Gould 2003: 242), as in historical research. In both senses, the network model is a useful tool for the description of connectedness within the context of an evolutionary view of knowledge development. The network is not a physical structural pathway for ideas pre-structured by a substantial element determining the fate of knowledge, but is as real as a map. It is an abstract representation of the selection and transfer of ideas.

In *Science as a Process*, Hull (2001 [1988]: 434) listed several qualities of the selection and transfer of ideas in scientific systems that can be interpreted readily as characteristics of a network as the structural pathway for the evolution of knowledge, with scientists serving as the "vehicles" for knowledge elements ("replicators"). Firstly, nodes and links may be appropriate representations of the idea that "progress in science occurs by means of recombinations" (Hull 2001 [1988]: 434) of existing ideas. The recombination of ideas in a network is symbolised by nodes representing ideas or scientists (as vehicles of ideas), and links representing the selection of combination of ideas carried by the vehicles. Secondly, a network representation allows for the symbolisation of "cross-lineage borrowing" of ideas (Hull 2001 [1988]: 450). An innovation may have multiple origins and can be transferred horizontally and vertically, which can be better represented in a network than, for example, in a bifurcating depiction of knowledge development. Thirdly, the conceptual kinship (Hull 2001 [1988]: 435) of different ideas can be displayed in a network. Common links can be used to symbolise scientists (as vehicles of ideas) who share ideas or elements of concepts with identical descent. Finally, different combinations of ideas resulting in the evolution of knowledge are the results of selection processes, which can be described suitably by a network. If selection (e.g. of ideas that are transmitted) is seen as an "interplay between replication and interaction" (Hull 2001 [1988]: 436f.), links between nodes symbolise positive selection and the exclusion of certain nodes in a diachronic perspective represents negative selection.

In the evolution of science, the exchange of ideas between network clusters perceived as, for example, "disciplines", and the recombination of these ideas may be hypothesised to lead to what is perceived as scientific progress. Although evolution is by definition undirected, supposed progress may be the result of the selection of the fittest concepts from diverse ideas. In other words, the borrowing of ideas from other disciplines leads to greater diversity, which improves the chances of finding a fit concept. Some findings of the network theory established in sociology relate extremely well to theories of innovations in science and technology. For example, Mark Granovetter's "strength of weak ties" hypothesis highlights that a small number of nodes in some network structures may have few links, but that these links may serve as bridges between clusters. Thus, these nodes are valuable elements because they link network clusters to one another, allowing for information exchange between clusters that would not have had contact without the respective nodes.[8] They serve as so-called "brokers", an intuitively understandable description of nodes representing social actors or elements of ideas. Several network analyses (with emphases on economics) have described "the diffusion of innovations" as such a networking process (Easley and Kleinberg 2010: 498).[9] At the same time, however, a highly interconnected idea might have a "selection advantage" because it is less susceptible to isolation following ruptures of connections. Other links can take up the roles of broken, deleted, or partitioned connections. Bearman et al. (2002: 66) have emphasised this point convincingly in arguing that even events reconstructed by an historian can be displayed in a network structure, because only connected events result in a meaningful historical event sequence. If the deletion of a link results in partition, the respective event might be interpreted as pure coincidence. Transferring this concept to networks of ideas would mean that ideas with very few links to other ideas might be forgotten quickly.

To describe the connectedness of ideas, the common descent of a thought from one origin and the selection of ideas from a diverse set, the deconstruction of broad scientific concepts into their elements is necessary, just as the identification of genes constituting a phenotype is necessary for the reconstruction of biological relations. One way of abstracting individual elements from a scientific concept is to apply frame theory, an approach borrowed from the cognitive sciences. Andersen, Barker, and Chen (2006) showed convincingly that this theory is a powerful tool for the dissection of concepts and analysis of the fate of their elements from a diachronic perspective. They focused on "The Cognitive Structure of Scientific Revolutions" in anatomising concepts to analyse, for example, Thomas Kuhn's concept of incommensurability. Inherent in this approach is also a very feasible method to describe the evolution of knowledge in the form of interconnected elements of ideas (i.e. Hull's replicators). The underlying idea of this approach is that semantic (also known as conceptual) knowledge forms a central basis for the use of language to describe solid facts and abstract terms (Klein 1999).[10] An essential aspect of semantic knowledge is the ability to categorise. Early cognitive-scientific approaches addressing the categorical structuring of semantic content used feature lists, which enable the definition of a distinct term by compiling its characteristic features (Rosch et al. 1976). These models were unidimensional and lacked flexibility, in contrast to approaches using frames that originated from a systematically networked structure of object / concept characteristics. Frames focus on the hierarchical order of characteristics that define a certain term (Barsalou 1992).[11] They are stereotypical and empirically founded structural formats for various forms and fields of

---

8   On Granovetter's hypothesis from the 1960s and further centrality measures in network analyses, see Easley and Kleinberg (2010: 43–47).

9   Coleman et al. (1957) published a path-breaking yet classical study using this approach. Collaboration networks in science have also been examined using citation analyses; see among others Bordons and Gómez (2000).

10   The following ideas have been outlined previously in German in Fangerau et al. (2009).

11   Marvin Minsky introduced the term "frame" in artificial intelligence research in his seminal study "A framework for representing knowledge" (1974). See also Minsky (1990).

knowledge (Minsky 1985: 244). As an entity, each frame is comparable to a concept. Conceptual knowledge is represented by the combination (and, from an evolutionary perspective, recombination) of information elements.[12]

A frame describes an object on the basis of general *attributes* to which specific *values* can be assigned (Barsalou 1992: 29–44). For example, numerous attributes can be assigned to the medical diagnostic concept of diabetes, such as the amount, colour, and taste of urine. Certain values are assigned to these attributes, according to an actual urine type (e.g. polyuria, oliguria, anuria; light, dark, sweet, salty). These values are subordinated to the attributes and represent a possibility within the attribute-value set. As such, they can form additional frames (e.g. sweet as a type of taste) or belong to higher-order frames (e.g. diabetes as a kind of disorder represented in urine). Frame analysis, understood as a diachronic network analysis of nodes and links representing elements of ideas and their transmission, enables the assignment of attributes and values in the course of temporal changes and thus describes phases of transition from one concept to another. The combination, selection, transfer, and recombination of elements of concepts to new concepts can be described in the form of interlinked attribute-value sets. Retrospectively, a researcher can determine whether transfer led to successful (fit) or unsuccessful recombinations, and whether selection blocked insight (e.g. by linguistic incompatibility), promoted it (e.g. by epistemologically sharper terminology), or even enabled new scientific approaches. Basically, logical breaks and inconsistencies in attribute-value constellations, which are the result of new empirical findings or reconfigurations of idea elements, result in conceptual shifts and, thus, evolution of knowledge.[13] The trans-temporal interconnections between attributes and values of concepts that stand for successful recombinations ultimately characterise a conceptual change, or what is seen as "progress".

However, one should not forget that the resulting structure of relationships among elements of ideas is a function of the underlying selection and transfer processes, not of their origin.[14] People, i.e. scientists, decide, which ideas or elements thereof they want to include in their network of ideas and which they want to discard. Hull sees scientists as "essential links in conceptual replication systems" (Hull 2001 [1988]: 447), who expect explicit or implicit credit for ideas or their transfer. They accept and select ideas for replication that they recognise as valuable or reasonable to be transferred. This social element of the system of science can also be depicted in networks. In a description of scientific development as a collective action, the abstract idea of a network to describe the evolution of knowledge becomes very concrete at this point.

---

12 In cognitivism, concepts, although universally determined, are understood as individual mental units (Strauß 1996: 42). I thank Michael Martin for raising this point and referring me to the relevant literature.

13 Andersen, Barker, and Chen (2006) have shown that frames can be used productively for the analysis of thought-style shifts on a conceptual level. They pointed to hierarchical fractures in frames through the introduction of novel attributes, values, or constraints that force a conceptual reorientation.

14 Gould (2003: 261) made a similar comment on other network data.

MAPPING THE SOCIAL STRUCTURE OF SCIENCE:
NETWORKS OF SELECTION AND TRANSMISSION[15]

In the history of science, so-called social constructivist theories have been used to interpret apparently objective scientific facts as products of the social conditions of research contexts. From this perspective, the production of knowledge gains the status of science as an organised practice only if not only individuals, but also collectives, believe in the prevalent methods of knowledge production and in the resulting scientific products (i.e. hypotheses, ideas, descriptions, new practices).[16] This social view of the establishment, implementation, and perpetuation of scientific theories, proposed by authors such as Kuhn (1962) in his path-breaking work on the structure of scientific revolutions and Latour (2005) in his far-reaching presentation of the actor-network approach, had already been put forward by Ludwik Fleck in the 1930s. Fleck (1979) described the "Genesis and development of a scientific fact" as a collective process and presented a model of how interactions among researchers, who form a thought collective, foster the creation of facts through negotiations of methods, hypotheses, and the validity of theories. With reference to Hans Vaihinger's "philosophy of 'as if'" (1924), the philosopher Arnold Kowaleski (1986 [1932]) proposed that scientific reasoning and the production of knowledge purposefully lead to fictions that are necessary for further development of the respective knowledge or useful on a practical (and methodological) level. In Kowalewski's view, in an environment of equally correct and/or acceptable fictions, only the collective recognition of certain fictions in a "community of ideas" ("*Ideengemeinschaft*") would lead to their implementation (Kowalewski 1986 [1932]). From an evolutionary standpoint, Kowalewski's ideas can be interpreted as proposing that ideas are selected and subsequently transferred in collective (networking) actions.

Reconstructing the processes of idea selection and transfer on the level of the scientific literature has a long tradition as bibliometrics in the information sciences.[17] In citation and/or co-citation analyses, articles published by authors serve as surrogate parameters for the ideas represented therein and for the authors as bearers of these ideas. Publications serve as an important substratum of accepted knowledge that is to be transferred. Citing and being cited in publications can be interpreted as recognising or being recognised as the result of a selection process. A network can be constructed through the examination of citations of authors in a corpus of literature and can be perceived as a snapshot of knowledge selection. By adding a temporal level, the transfer of selected elements can be represented, aiding the visualisation of an evolutionary process. Citations (as surrogate parameters for scientists and their ideas) that crosslink texts can be considered to represent an intellectual network and a symbolic social network constructed strategically by au-

---

15   Some of the following thoughts have been published elsewhere with a different focus in German (Fangerau 2009a; Fangerau 2010b).

16   Jan Golinski (2005) has provided a summary account of these views.

17   For an overview, see the contributions published in a festschrift for Eugene Garfield (Cronin and Atkins 2000).

thors, which differs markedly from (unintentional and undirected) natural selection (see above). Citation analysis of interlinked texts goes well beyond the examination of personal contacts by broadening the scope to thoughts that authors cited together might have shared. Works cited together can be viewed as the "intellectual base" of current knowledge, the starting point for further evolution in a research field. The citing works represent the research frontier or, from an evolutionary perspective, "evolving" variations of knowledge (Chen 2003a; Chen 2003b; Chen 2004; Persson 1994: 31). Scientific literature is published in journals that follow specific norms. Thus, this journal based communication among scientists can be considered to be the formal communication system in science. Therefore, I consider thought collectives reconstructed through the analysis of citation patterns to be "formal thought collectives" (Fangerau 2009a; Fangerau 2010b).

The analysis of formal thought collectives is of particular value in the analysis of interdisciplinary transfer and the mapping of scientific fields (Hull's "demic structures"; see above). However, they are less valuable in the examination of networks including people who do not participate in the formal communication system of science. This situation may exist in the case of the examination of the influx of tacit knowledge and its influence on the evolution of scientific knowledge in the history of medicine or technology.[18] For example, innovations in tissue engineering have been described as the results of co-evolution of scientific and technological networks characterised by "co-mingling" through founding, consulting, or advising, rather than by co-publishing or citing (Murray 2002).

In such cases, historical social network analyses that examine direct contacts between actors seem to be advantageous because they can aid the historical investigation not only of formal scientific links, but also of informal links. They enable the exploration of links that may have been essential for intercultural transfer based more on skills and mutual and implicit knowledge than on formal scientific transfer, which represents only explicit knowledge. Furthermore, they may aid the detection of reasons for the selection and selective transfer of specific ideas or, more concretely, the citation of authors as representatives of ideas. Of course, extended social network analyses examining contacts and the content of information possibly exchanged are necessary in transfer studies. Unfortunately, historians, unlike sociologists, usually do not have the opportunity to systematically question protagonists. Thus, they must rely on other sources to gather information about the existence, quality, and content of social contacts. Possible sources for such an endeavour are publications, autobiographies, institutional bonds, familial contacts, and – a very useful special source – correspondence (Steinke 2004).

18 For example, Knudsen (2003) proposed that tacit knowledge be considered a very important factor in evolutionary epistemology and offered what he called a "neo-Darwinian model of science" for the description of knowledge evolution. He characterised it as based on emulative selection that combined a "Darwinian replication of implicit scientific knowledge" with "Lamarckian replication of explicit scientific knowledge". He described Darwinian evolution as characterised by the "replication of the cultural code with recombination and random component" and Lamarckian evolution as changes in the replicators' code by "environmental or idiosyncratic stimuli", followed by the "replication of the altered code" (Knudsen 2003: 110f.).

In addition to qualitative analyses, semi-quantitative examination of the network of correspondents and the individuals mentioned in letters may be particularly valuable.[19] Presuming that the occurrence of names in correspondence, similar to citations in texts, hints at the writers' or addressees' engagements with these individuals, the networks evolving from such an analysis reflect direct and indirect social contacts and, like citations, intellectual networks. Individuals mentioned in correspondence can have one or more of four character traits: 1) they can have formal and informal personal contacts with one or both correspondents without knowing them face to face, 2) they can have direct social relationships with one or both correspondents, 3) they can have been subject of intellectual engagement with them by one or both correspondents, or 4) they have no relationship with the correspondents but have influenced their reasoning to the extent that they are mentioned in letters (e.g. politicians). Because these socio-intellectual networks also represent thought collectives and are reconstructed from informal communication rather than scientific journals, I consider them to be "informal thought collectives" (Fangerau 2009a; Fangerau 2010b).

To summarise, networks of scientists can be analysed using two levels of representation. Whereas formal citation networks can be compared to a "mind map" of scientific ideas, informal social networks represent personal and intellectual contacts and, especially in transfer analyses, assumed knowledge exchange through contact. Methods of citation network analyses from the information sciences and those of social network analyses from the social sciences have been implemented successfully in history.[20] Nevertheless, they are underrepresented as an element of the historical investigation of evolutionary epistemology, although citation studies have been used, for example, to visualise scientific paradigms and their development (Chen 2003a; Chen 2003b; Chen 2004). One reason for this underrepresentation might be that the formal character of citation analyses underestimates social motivations of citations in networks of recognition; in other words, citation analyses underestimate the authors' motivations to select or neglect specific peer authors, although they might have been included in a publication intended to transfer knowledge (e.g. from one generation of scientists to another or in the form of lateral transfer to peers). Scientists are largely characterised within networks by the aim to gain recognition and the power to evaluate and select peers' works. The power of recognition and associated self-constitution as a driving force in scientific network building may be illustrated by the following theoretical considerations, followed by an empirical example of the purposefully motivated "evolutionary" selection of knowledge. The example also shows the double character and limits of citations as apparently simple markers of recognition.

19 On network studies see for example Dauser (2008); Mauelshagen (2003); Mücke and Schnalke (2009); Pearl (1984); Rusnock (1999); Steinke and Stuber (2004).
20 On historical applications of social network analysis methods, see for example Garfield (1973), Garfield et al. (2002), Wetherell (1998), Gould (2003), Reitmeyer and Marx (2010), Lemercier (2011; 2012).

RECOGNITION AS A DRIVING FORCE IN THE SELECTION
AND TRANSFER OF IDEAS

Authors such as Robert Merton (1973) and Richard Whitley have examined the problem of collectively selecting knowledge by recognising it as relevant and correct with respect to the questions addressed when describing "the intellectual and social organization" of the sciences as a "reputational system" (Whitley 1984). These authors have added the social analytic category of the motivational aspect of "reputation" to the investigation of organisation and communication in the analysis of science. According to Whitley (1984), the self-defined goal to create, find, or describe new knowledge goes hand in hand with insecurity about the exact goal of a scientific endeavour. In addition to methodological constraints, science has a hermeneutic element that Whitley described as "task uncertainty". This insecurity requires repeated reconfirmation with colleagues and peers, which secures that a researcher is working toward a common goal of his thought collective using methods accepted by this collective and that this work has the chance of garnering recognition (Whitley 1984).

Whereas reputation is similar to social capital as defined by Bourdieu (1983), recognition is a more dimensional currency.[21] The concept of recognition in this case encompasses two semantic fields that I believe to be hardly divisible in the context of science: the acknowledgment of an idea, paradigm, or theory as "possibly true", "plausible", or fitting empirical reality; and respect or appreciation that builds a scientist's reputation. In both cases, recognition indicates a relationship and thus structures the selection and transfer of ideas in science or "scientific knowledge". The reconstruction of flows of recognition and the analysis of external influences on these flows yields a multidimensional relational structure that can be represented in a network format. As in Latour's actor network theory (Latour 2005), different kinds of actor can be represented as influential in such a network. However, when the research addresses from a historical perspective questions of knowledge selection and transfer, a focus on prosopographical approaches or, at least, on surrogate parameters for scientists and their interactions, as far as these can be reconstructed from historical sources such as correspondence or literature, would be useful. Correspondence and scientific publications (e. g. journal articles, books) are valuable "vehicles" of ideas and, from a historical perspective, more concentrated replicators of concepts than the scientist who produced them, because "the scientist", who plays many roles in his or her life, is obviously much more of an abstraction from an idea than are his written products.[22] Citations are a valuable element of written products for the reconstruction of evolution, which includes scientists and their ideas. Tracing scientists' reasons for citing others provides some insight into the roles of recognition and self-constitution in the evolution of knowledge.

Scientists cite (and thereby recognise) other authors because the scientific practice of citing ideally shows their ability to work according to collectively accepted

21   On the concept of recognition in a much more complex understanding than used here see above
     all Honneth (1996)
22   Of course, a person has many more facet than his or her knowledge production.

methods (by indicating their familiarity with the state of research), reveals their reference points, and objectifies their ideas. From a more realistic and critical perspective, citation rates are measures of authors' general motives to be concerned with other scientists and their publications and, further, to articulate this intellectual engagement (Stock 1985: 314). In this sense, a citation captures the multifaceted elements of recognition and associated selection and transfer described above.

Scientific authors tend to cite themselves and their colleagues more than once in a series of publications. Thus, an "invisible college" (Crane 1972) can be reconstructed from the intertextual network and structured using quantitative (and qualitative) methods to define central and peripheral bearers of ideas (interactors). Those who are cited more often are attributed with a high degree of positive or negative recognition (opponents can honour and recognise their counterparts as respected scientists worth citation or "negatively recognise" another scientist to display the implausibility ("unfitness") of an idea). The pattern of re-citations in each author's corpus is unique; Howard White called this pattern "citation identity" and compared it to a fingerprint (White 2000; White 2001). An author's "citation identity" must be differentiated from his or her "citation image", which can be constituted by examining the citation contexts in which his or her works are cited. Both terms suggest how processes of recognition, and thus selection and transmission, can be reconstructed through the analysis of identities and images.

At this point, the limitations of the narrow understanding of a citation as a representation of an interactor's recognition of an idea as true or false become evident. Moreover, the dual nature of the concept of recognition described above becomes apparent because citations are much more than statements about which ideas an author is currently following. They additionally serve the purposes of establishing ties with peers by seeking allies through positive citations (selecting their ideas) and of making other interactors and their ideas redundant through negative citations or neglect of their works (Bavelas 1978). Finally, self-citation has the positive effect of spreading one's ideas or increasing their recognition among peers in the long term through repetition.[23] Thus, through the citation practice, an author participates in selection and transmission processes driven not only by the identification and dissemination of information considered to be relevant (Cronin 1981), but also by social components associated with recognition and reputation (Cozzens 1989). As Cronin and Shaw (2002) have noted, Bourdieu (1992: 43, 143) viewed citation rates as surrogate parameters for symbolic capital in science and Cary Nelson (1997: 39) compared citations to "academia's version of applause".

Combined network analyses of formal and informal thought collectives aid the detection of these social selective forces in such collectives by identifying intellectual links 1) between individuals who know each other personally, 2) between authors with no personal relationship who share a field of expertise, and 3) between authors and their authorities (who may be deceased). However, citations can be made for more than one reason, ranging from the pure identification and dissemina-

---

23   On current practices of self-citation that might be conferred to historical studies, see Falagas and Kavvadia (2006), Glanzel et al. (2006).

tion of information to a reflection of the social component of scientific communication within the (positive or negative) reputational system of science.

This multidimensionality and the motivational aspects of selecting and transmitting knowledge by citing can be illustrated with the historical example of the citation norms and practice of the German-American physiologist Jacques Loeb (1859–1924), who addressed the problem of recognition in science multiple times in correspondence. Against the given theoretical framework, his example may serve as a test of the hypothesis that recognition plays an important role in the selection and transfer of ideas as well as in the self-constitution of researchers as scientists. Loeb was an authority in science around 1900, as a representative of reductionist biomedicine. He worked as a biomedical researcher, engaged in the politics of science, founded a journal and handbook series, and tried to foster international scholarship. His views are revealing because they illustrate how the evolution of knowledge is influenced by scientists' motivations to be recognised in networks and how citation practices evolved as a scientific methodological norm.[24]

THE EXAMPLE OF JACQUES LOEB[25]

Loeb was a German physiologist who immigrated to the United States after marrying an American woman. There, he became a world-renowned scientific celebrity after publishing the results of experiments in "artificial parthenogenesis", in which he successfully induced development in sea urchin eggs without sperm by changing the electrolyte balance in their surrounding fluids (Fangerau 2010b; Pauly 1987). During his lifetime, he was portrayed as an archetypical reductionist physiologist who tried to explain all life phenomena on a physicochemical basis (Fangerau 2006). Even Mark Twain (1835–1910) referred to his work; in 1905 (Brodwin 1995: 242), he answered a sceptical comment on Loeb's research in the *New York Times* in favour of Loeb, stating that "a consensus of opinion among biologists would show that he [Loeb] is voted rather as a man of lively imagination than an inerrant investigator of natural phenomena" (Twain 1923: 304). In his reply, Twain – ironically in contradiction to the constructivist view of the evolution of science presented here – warned scientists that looking for consensus would hinder innovation and new discoveries in science (Twain 1923).

Loeb considered himself an entrepreneur of purely reductionist scientific medicine. He tried to help in the selection and transfer of a biomedical model in the evolution of medicine. When he commenced work at the Rockefeller Institute for Medical Research in 1910, he claimed that experimental biology should become the basis of medicine. At the same time, he saw the need to campaign for recognition for this idea. He feared "… that the Medical Schools in this country are ready for the new departure […] The medical public at large does not yet fully see the bearing

---

24 This, of course, is an example of the evolution of science as an endeavour or, in other words, the evolution of scientific practice (not ideas).

25 The following section and the quotations from archival material have been published in German in Fangerau (2010b).

of the new Science of Experim. Biol. (in the sense in which I understand it) on Medicine."[26]

One method of pushing the evolution of medicine in this direction (again different from natural selection) was directing recognition for his model by selecting and transmitting scientific citations in his and other authors' works. Loeb's invited review of a manuscript submitted to the *Journal of Experimental Zoology* serves as an example reflecting the polar goals of citing and reflecting citations' relation to recognition as a driving force in the evolution of science. Loeb wrote, "I return the manuscript of Mr. Lowe, [...] the man, as is usual with our young writers, does not know the literature. To give an example, the more important part of his paper deals with the influence of various potassium salts, especially the influence of the anion. He has overlooked the fact that the last year Cattell and I published a long paper on this very subject [...] If I may make a suggestion, you return the paper to Mr. Lowe, requesting him to look up this missing literature and to utilize it in his paper."[27]

This quotation provides empirical evidence for the previously discussed theoretical considerations about citations, recognition, and selection. Loeb gave two reasons for the absence of certain citations in Lowe's manuscript. First, he criticised Lowe for failing to document his knowledge of the methodological and conceptual state of research. Or – in the language of evolution – he, as an interactor, did not document that his thinking was based on the variant of knowledge that Loeb considered to best fit "reality". [28] Second, Loeb criticised Lowe's failure to cite his work, which he interpreted – in addition to Lowe's missing knowledge – as a refusal of recognition. In his struggle to make his research relevant, recognised, selected, and transmitted, Loeb negatively selected Lowe's paper and suggested that the journal not publish it in its present form. Loeb probably would have abstained from comparing scientific activity or his actions with evolutionary ideas because he considered "evolution" to be an unscientific grand theory without meaning (Fangerau 2010b: 219). As he wrote in a letter to Ernst Mach (1838–1916), being cited was – after experiences of being in the periphery of science during his early career in Germany –"a great gratification"[29] for him. Nevertheless, his motivation to receive gratification and recognition can be retrospectively linked to ideas of evolutionary selection and transmission in science.

The topos of recognition through citation played a major role in Loeb's correspondence with various scientists; particularly, he treated priority disputes – with priority representing the highest form of reputation and recognition of new ideas that survive selection because they are "fit" – very sensibly. For example, when

26   Loeb to Simon Flexner (1863–1946), cited in Osterhout (1928: 328).
27   Loeb to Harrison 17.07.1916, Archives of the Library of Congress, Loeb Papers (LOC) (Fangerau 2010b: 128).
28   Loeb wrote similar comments about other authors, for example Loeb to Harrison 05.01.1924, LOC: "The papers by Miss Collett show that she is not familiar with the recent literature on the subject" (Fangerau 2010b: 128).
29   Loeb to Mach 21.02.1903, Archiv des Deutschen Museums, München, Nachlas Ernst Mach (DMM) (Fangerau 2010b: 128): "Wilhelm Ostwald has cited my works repeatedly and this is a great gratification for me" (original in German, translation by HF). On Loeb's disappointing experiences as a young scholar in Germany, see Pauly (1987).

Paul Ehrlich (1854–1915) complained about being unrecognised in a publication from Loeb's laboratory,[30] Loeb was quick to write a clarifying answer to Ehrlich stating: "It is self-evident that I will, as soon as an opportunity occurs, fix this issue … nothing is a greater obstacle for the right development of a scientific thought than if the real author and inventor is put aside; and I have always regarded it as my duty, to stand up for the correct presentation of the case".[31]

Loeb himself felt that his German counterpart and competitor Max Verworn (1863–1921) referred to his research particularly insufficiently.[32] Both researchers were striving to establish a "general physiology" on the basis of their personal works.[33] In contrast to Loeb, Verworn did not understand this general physiology as reductionist and comparative on the basis of physics and chemistry. Rather, he wished to promote a holistically oriented cellular physiology (Verworn 1895: 50 ff.). Loeb vehemently attacked this idea, but his major criticism was that Verworn did not cite his (Loeb's) works. In a polemic in the *Archiv für die gesamte Physiologie*, he stated: "It is strange by which means authors and their addresses are eliminated by these cellular-physiologists. The number of my papers on general physiology for example is higher than the one of Verworn. Nevertheless, my observations are not mentioned in his handbook, although their results were welcomed by him in his conclusions. The 'cell-state' is a mere phrase, which is scientifically worthless. But this phrase is enough for Verworn to eliminate my works, which are in the sense of Claude Bernard contributions to general physiology. [Wilhelm] Roux has already complained about Verworn's literary peculiarities, and he has called him in plain terms a copyist. Verworn felt hurt by this, but it seems to me, that he should better leave the feeling of being hurt to those who are disposed of the fruits of their labour by his cellular-physiologic sophism" (Loeb 1897–1898).

Just before the young zoologist (and later colloid chemist) Wolfgang Ostwald (1883–1943), the son of the Nobel Prize winner Wilhelm Ostwald (1853–1932), came to Berkeley to work as Loeb's research-assistant, they exchanged letters about "Verworn's manner" of not citing peers.[34] Both felt that the failure to cite authors

30 Ehrlich to Loeb 03.07.1906, LOC (Fangerau 2010b: 128). The letter by Ehrlich reveals another important factor in the psychology of citing. He ends by recognising Loeb as an international authority in science and expects to be recognized in return: "… I was very disappointed to see, that in these works about lipoids only Hans Mayer and Overton and in the newest work by Dr. Brailsford Robertson in the Journal of Biological Chemistry only Lowell and Hamburger are cited, while my name is not mentioned at all. I am even more disappointed, because these are works from your institute, which receive highest attention all over the world" (original in German, translation by HF).

31 Loeb to Ehrlich 20.07.1906, LOC (Fangerau 2010b: 129). Original in German, translation by HF.

32 Loeb to Mach 17.05.1897, DMM (Fangerau 2010b: 129).

33 Philipp Pauly called Max Verworn Loeb's "mirror image" (Pauly 1987: 84). On Loeb and Verworn, see also Fangerau (2010a), Fangerau (2012: 228–32).

34 Wolfgang Ostwald to Loeb 29.04.1903, LOC (Fangerau 2010b: 130). Ironically, after 1914 and the beginning of World War I, the same Ostwald angered Loeb because Loeb felt that Ostwald had not directly cited or recognised him sufficiently (Fangerau 2010). On a more general structural level than citations, after the end of the war Loeb helped to disseminate his reductionist concept of science in Europe by financially supporting only selected scientists and scientific

was acceptable only if concepts introduced into science had been established on the level of general knowledge.[35] This notion reveals a further link to an evolutionary understanding of science: transfer and selection seem to happen only on the level of new variations. Variations established under certain conditions show a greater tendency to survive because they are not under pressure to adapt. A scientific revolution (in Kuhn's terms) or an extreme change in the thought environment (in evolutionary terms) is necessary to challenge established ways of thinking. Only then must authors be brought in again as representatives, surrogate parameters, or proxies for ideas.

Scientific recognition, in its double meaning of accepting ideas and honouring scientists for their work in the sense of reputation, played an important role in Loeb's practice of scientific communication. His exchange with Ehrlich suggests that his stance was a fundamental position of scientists who wanted to foster the survival of their ideas in the network of a thought collective. Additionally, analysis of his correspondence clearly reveals that Loeb constituted his scientific self on the basis of being recognised and recognising.

DISCUSSION AND OUTLOOK

Niklas Luhmann (1998: 575) argued that the combination of theories might lead to fruitful questions in history. In other fields, contact, the exchange of ideas, and their lateral transfer across disciplinary borders may also have innovative effects by increasing diversity through cross fertilisation if at least some elements are compatible. Loeb, for example, combined ideas from organic chemistry, psychology, and experimental physiology to achieve some of his results, such as the description of "artificial parthenogenesis" or "heteromorphosis". His interdisciplinary approach – including formal and informal networks – around 1900 resulted in the formulation of his research programme of general physiology.[36] Similarly, ideas have been exchanged successfully between linguistics and biology for the last 150 years, yielding overarching or at least similar concepts of the evolution of their research objects (language and organisms).[37]

In this text, I have made the slightly self-reflexive attempt to select and recombine elements of network and evolutionary theories to develop a tool for the de-

literature. He wanted to fuel the evolution of his physicochemical approach to physiology, instead of an emergence-oriented holistic approach. For details, see Fangerau (2009b). A similar discussion with a different emphasis is presented in Fangerau (2007).

35   Wolfgang Ostwald to Loeb 16.05.1904, LOC (Fangerau 2010b: 130). This acceptance of not citing authors whose ideas have become common knowledge might be linked to the idea of Knudson, that explicit knowledge must be supported by implicit knowledge (Knudsen 2003). Once ideas are fully accepted they might transgress the spheres of explicit knowledge and become implicit knowledge.

36   A network description of the disciplines from which he borrowed ideas from is presented in Fangerau (2010b).

37   On interdisciplinary networks as a basis for innovation, see Andersen (this volume); on the historical intertwinement of linguistics and biology, see Kressing and Krischel (this volume).

scription of evolution in science as a networking process. I consider this approach to be reasonable because a network represents the kinship of ideas and/or methods or thought styles by displaying descent from a common origin, as well as social aspects of science, such as its demic structure or organisation in disciplines, respectively. When applied to the evolution of ideas in this sense, a network represents more than social ties. Rather, it is a map of connections between actors and/or elements of ideas (sometimes represented by authors) which – if a temporal dimension is included – is also a map of the connectedness of ideas resulting from selection and transmission. Nodes in the network represent elements of ideas or scientists as carriers of ideas, and links between nodes illustrate relationships of recognition that might have been influenced by authority, reputation, trust, personal relationships, intellectual communication, practical transfer, and other forms of scientific currency. On the cognitive level, Andersen, Barker and Chen (2006) showed how the recognition of a concept as plausible and logically consistent functions as a selective factor in so-called frames (see above); on the social level, recognition serves as a selective factor in its double meaning as recognising a scientist's statement as plausible and recognising his or her reputation.

In this way, science as a social practice has many dimensions ranging from the plausibility of an idea/theory to the authority of its carrier to the carrier's credibility, which is composed in part of his or her reputation in social and cognitive collectives of scientists. The network model seems to be helpful in capturing the relational aspects of recognition, not only in bilateral relationships between scientists, but also in a multi-relational way with regard to thought collectives, their ideas, and their diachronic evolution. The network approach is more than a metaphor in this case. It is also more than a sociological method applied to the history of science. It is a theory. The coding of data in the form of frames, citations, or the occurrence of scientists' names in correspondence is an abstraction from reality that is required by the theory that their interconnectedness results in something – in this analysis, the evolution of science. Thus, network theory constrains the method of reading historical data in a specific manner, and the general possibility of methodologically producing these data from historical material seems to justify the theory.[38]

However, one must be aware 1) of the importance of carefully defining and distinguishing data that one intends to examine from a relational perspective,[39] 2) of the time-consuming nature of collecting these relational data, and 3) that a network representing the evolution of a specific idea will never be complete. Although the survival in texts of at least some hints of relations is a characteristic element of the history of science, the complete reconstruction even of an ego network of one scientist is hardly possible. Finally, one must be aware of the old chicken-and-egg

---

38 Following Larry Laudan (1984), Hull (2001 [1988]: 447 f.) described this relationship between theory and method in general (not in relation to network analysis). On the problem of whether network analysis is a metaphor, a method, or a theory, see also Bögenhold and Marschall (2010a; 2010b), Gould (2003).

39 As Claire Lemercier (2011: 6) put it, "Choices in 'boundary specification' (whom do we observe? which ties among them? at what time(s)?) heavily constrain the sort of questions that can be analyzed by network analysis". See also Lemercier (2012).

problem. Historical network analyses seeking to reconstruct the evolution of ideas examine the fates of elements of ideas in the world, not necessarily their "generation" *ab ovo*. The abstract links connecting all of these facets might be captured by the idea of recognition in science. This approach, however, entails the acceptance of the constructivist view that processes of transfer and selection follow the more or less rational choices of actors, guided by the logic of their disciplines.

The advantage of the historical reconstruction of networks of idea exchange is that it enables an empirical approach based on historical data. Historical network analyses not only follow the heuristic aim of providing an analytical framework for qualitative analyses of communication in science, but can also help to provide both a reconstruction of recognition flows (on a direct level) and information about the evolution of the structures of scientific thought collectives and the evolution of ideas (on a more remote, abstract level), if the nodes represent vehicles or elements of ideas. Above all, however, network representations aid the analysis of selection processes that ultimately lead to the transmission of ideas. Scientific selection is fuelled by recognition in all its meanings. Howard White (2008) aptly described this characteristic of networks in terms of "identity and control". In concordance with the proposed concept of recognition and self-constitution as driving forces of scientific evolution, White proposed that identities (e. g. of scientists) are developed only within network-like social structures. These social structures can be interpreted as organisational structures – in scientific contexts, thought collectives or disciplines that are formed and controlled only by the sustainment of relationships based on recognition among the actors involved (White 2008).[40,]

Overall, if the general idea of explaining scientific evolution as the evolution of ideas and scientific structures in the form of networks is accepted, then recognition and self-constitution could reasonably be considered to be driving forces of selection and transfer in science. Conversely, if the social structure of science as a constituent of scientific practice is accepted, then the analysis of processes of scientific selection as social processes that can be displayed in networks would be reasonable. The strength of the historical reconstruction of such networks is that it allows for the analysis of the selection and transfer of ideas that led to shifts in relationships of recognition. From an evolutionary standpoint, these shifts result in scientific "progress".

---

40   In a similar form, George Homans' (behaviorist) exchange theory, with its focus on interaction and norms (Homans 1958), might help to describe the evolution of ideas as the result of selection fostered by recognition. According to this theory, one could say that also scientific norms are reinforced by interaction and re-selection, and that norms dissolve through negative selection.

REFERENCES

Andersen, H., P. Barker & X. Chen (2006) The Cognitive Structure of Scientific Revolutions (Cambridge/New York: Cambridge University Press).

Barsalou, L.W. (1992) 'Frames, Concepts, and Conceptual Fields', in A. Lehrer & E.F. Kittay (eds), Frames, Fields, and Contrasts (Hillsday): 21–74.

Bavelas, J.B. (1978) 'The social psychology of citations', Canadian psychological review 19(2): 158–63.

Bearman, P., J. Moody & R. Faris (2002) 'Networks and History', Complexity 8: 61–71.

Bögenhold, D. & J. Marschall (2010a) 'Metapher, Methode, Theorie. Netzwerkforschung in der Wirtschaftssoziologie', in C. Stegbauer (ed), Netzwerkanalyse und Netzwerktheorie. Ein neues Paradigma in den Sozialwissenschaften (Wiesbaden: VS Verlag für Sozialwissenschaften): 387–400.

Bögenhold, D. & J. Marschall (2010b) 'Weder Methode noch Metapher. Zum Theorieanspruch der Netzwerkanalyse bis in die 1980er Jahre', in C. Stegbauer & R. Häußling (eds), Handbuch Netzwerkforschung (Wiesbaden: VS Verlag für Sozialwissenschaften): 281–9.

Bordons, M. & I. Gómez (2000) 'Collaboration Networks in Science', in B. Cronin & H.B. Atkins (eds), The Web of Knowledge. A Festschrift in Honor of Eugene Garfield (Medford, New Jersey: Information Today): 197–213.

Bourdieu, P. (1983) 'Ökonomisches Kapital, kulturelles Kapital, soziales Kapital', in R. Kreckel (ed), Soziale Ungleichheiten (Soziale Welt, Sonderband 2) (Göttingen: Otto Schwartz & Co).

Bourdieu, P. (1992) Homo Academicus (Frankfurt: Suhrkamp).

Brodwin, S. (1995) 'Mark Twain's Theology: The Gods of a Brevet Presbyterian', in F.G. Robinson (ed), The Cambridge Companion to Mark Twain (Cambridge: Cambridge University Press): 220–48.

Campbell, D.T. (1974) 'Evolutionary Epistemology', in P.A. Schilpp (ed), The Philosophy of Karl Popper (La Salle, Illinois: Open Court): 413–63.

Chen, C. (2003a) Mapping scientific frontiers: the quest for knowledge visualization (London: Springer).

Chen, C. (2003b) 'Visualizing scientific paradigms: An introduction', Journal of the American Society for Information Science and Technology 54(5): 392–3.

Chen, Ch. (2004) 'Searching for intellectual turning points: progressive knowledge domain visualization', Proceedings of the National Academy of Science 101 Suppl 1: 5303–10.

Churchland, P.M. & C.A. Hooker (1985) Images of Science: Essays on Realism and Empiricism, with a Reply from Bas C. van Fraassen (Chicago: University of Chicago Press).

Coleman, J., E. Katz & H. Menzel (1957) 'The Diffusion of an Innovation Among Physicians', Sociometry 20(4): 253–70.

Collin, F. (2003) 'Evolutionary, Constructivist and Reflexive Models of Science', in H.S. Jensen, L.M. Richter & M.T. Vendelo (eds), The Evolution of Scientific Knowledge (Cheltenham: Edward Elgar): 57–78.

Cozzens, S.E. (1989) 'What Do Citations Count – the Rhetoric-First Model', Scientometrics 15(5–6): 437–47.

Crane, D. (1972) Invisible Colleges: Diffusion of Knowledge in Scientific Communities (Chicago: University of Chicago Press).

Cronin, B. (1981) 'The need for a theory of citing', Journal of Documentation 37(1): 16–24.

Cronin, B. & H. Barsky Atkins (2000) The Web of Knowledge: A Festschrift in Honor of Eugene Garfield (Medford, N.J.: Information Today).

Cronin, B. & D. Shaw (2002) 'Banking (on) Different Forms of Symbolic Capital', Journal of the American Society for Information Science and Technology 53(14): 1267–79.

Dauser, R. (2008) Wissen im Netz: Botanik und Pflanzentransfer in europäischen Korrespondenznetzen des 18. Jahrhunderts (Berlin: Akademie-Verlag).

Dear, P. (2012) 'Science is Dead; Long Live Science', Osiris 27: 37–35.

Easley, D. & J. Kleinberg (2010) Networks, Crowds, and Markets: Reasoning About a Highly Con-
    nected World (New York: Cambridge University Press).
Falagas, M.E. & P. Kavvadia (2006) '"Eigenlob": self-citation in biomedical journals', The FASEB
    Journal 20: 1039–42.
Fangerau, H. (2006) 'The Novel Arrowsmith, Paul de Kruif (1890–1971) and Jacques Loeb (1859–
    1924): A Literary Portrait of "Medical Science"', Medical Humanities 32: 82–7.
Fangerau, H. (2007) 'Biology and War', History and Philosophy of the Life Sciences 29(4): 395–
    427.
Fangerau, H. (2009a) 'Der Austausch von Wissen und die rekonstruktive Visualisierung formeller
    und informeller Denkkollektive', in H. Fangerau & T. Halling (eds), Netzwerke. Allgemeine
    Theorie oder Universalmetapher in den Wissenschaften? Ein transdisziplinärer Überblick
    (Bielefeld: Transcript): 215–46.
Fangerau, H. (2009b) 'From Mephistopheles to Iesajah: Jacques Loeb, Science and Modernism',
    Social Studies of Science 39: 229–56.
Fangerau, H. (2010a) 'Biologie und Erster Weltkrieg: 'General Physiology' und ihr Ursprung im
    Unfrieden', Verhandlungen zur Geschichte und Theorie der Biologie 15: 171–80.
Fangerau, H. (2010b) Spinning the Scientific Web: Jacques Loeb (1859–1924) und sein Programm
    einer internationalen biomedizinischen Grundlagenforschung (Berlin: Akademie Verlag).
Fangerau, H. (2012) 'Monism, Racial Hygiene, and National Socialism', in T. Weir (ed), Monism.
    Science, Philosophy, Religion, and the History of a Worldview (New York: Palgrave): 223–47.
Fangerau, H., M. Martin & R. Lindenberg (2009) 'Vernetztes Wissen: Kognitive Frames, neuronale
    Netze und ihre Anwendung im medizinhistorischen Diskurs', in H. Fangerau & T. Halling
    (eds), Netzwerke. Allgemeine Theorie oder Universalmetapher in den Wissenschaften? Ein
    transdisziplinärer Überblick (Bielefeld: Transcript): 29–48.
Fleck, L. (1979) Genesis and Development of a Scientific Fact (Chicago: University of Chicago
    Press).
Freeman, L.C. (2004) The Development of Social Network Analysis: A Study in the Sociology of
    Science (Vancouver, BC; North Charleston, S.C.: Empirical Press; BookSurge).
Garfield, E. (1973) 'Historiographs, Librarianship, and the History of Science', in C.H. Rawski (ed),
    Toward a Theory of Librarianship: Papers in Honor of Jesse Hauk Shera (Metuchen, NJ: Scare-
    crow Press): 380–402.
Garfield, E., A.I. Pudovkin & V.S. Istomin (2002) 'Algorithmic Citation-linked Historiography –
    Mapping the Literature of Science', Asist 2002: Proceedings of the 65th Asist Annual Meeting
    39: 14–24.
Glanzel, W., K. Debackere, B. Thijs & A. Schubert (2006) 'A concise review on the role of author
    self-citations in information science, bibliometrics and science policy', Scientometrics 67(2):
    263–77.
Golinski, J. (2005) Making Natural Knowledge: Constructivism and the History of Science (Chi-
    cago: University of Chicago Press).
Gould, R.V. (2003) 'Uses of Network Tools in Comparaive Historical Research', in J. Mahoney &
    D. Rueschemeyer (eds), Comparative Historical Analysis in the Social Sciences (Cambridge,
    New York: Cambridge University Press): 241–69.
Grantham, T.A. (2000) 'Evolutionary Epistemology, Social Epistemology, and the Demic Structure
    of Science', Biology and Philosophy 15: 443–63.
Hesse, M.B. (1974) The Structure of Scientific Inference (Berkeley: University of California Press).
Homans, G.C. (1958) 'Social Behavior as Exchange', American Journal of Sociology 63(6): 597–
    606.
Honneth, A. (1996) The Struggle for Recognition: the Moral Grammar of Social Conflicts (Cam-
    bridge, Mass.: MIT Press).
Hull, D.L. (2001 [1988]) Science as a Process: an Evolutionary Account of the Social and Concep-
    tual Development of Science (Chicago: University of Chicago Press).
Klein, J. (1999) ''Frame' als semantischer Theoriebegriff und als wissensdiagnostisches Instrumen-

tarium', in I. Pohl (ed), Interdisziplinarität und Methodenpluralismus in der Semantikforschung (Frankfurt: Peter Lang): 157–83.

Knudsen, T. (2003) 'A neo-Darwinian Model of Science', in H.S. Jensen, L.M. Richter & M.T. Vendelo (eds), The Evolution of Scientific Knowledge (Cheltenham: Edward Elgar): 79–119.

Kowalewski, A. (1986 [1932]) 'Die Haupteigenschaften der Philosophie des Als Ob', in A. Seidel (ed), Die Philosophie des Als Ob und das Leben. Festschrift zu Hans Vaihingers 80. Geburtstag. Neudruck der Ausgabe Berlin 1932 (Aalen: Scientia): 227–35.

Krischel, M., T. Halling & H. Fangerau (2012): 'Anerkennung in den Wissenschaften sichtbar machen: Wie die Bibliometrie durch die soziale Netzwerkanalyse neue Impulse erhält', Österreichische Zeitschrift für Geschichtswissenschaften 23: 179-206.

Kuhn, T.S. (1962) The Structure of Scientific Revolutions (Chicago: University of Chicago Press).

Latour, B. (2005) Reassembling the Social: An Introduction to Actor-Network-Theory (Oxford/New York: Oxford University Press).

Laudan, L. (1984) Science and Values: the Aims of Science and Their Role in Scientific Debate (Berkeley: University of California Press).

Lemercier, C. (2011) 'Formal network methods in history: why and how?', http://halshs.archives-ouvertes.fr/halshs-00521527, version 2 [last access 08.05.2013].

Lemercier, C. (2012) 'Formale Methoden der Netzwerkanalyse in den Geschichtswissenschaften: Warum und Wie?', Österreichische Zeitschrift für Geschichtswissenschaft 23: 16–41.

Loasby, B.J. (2002) 'The Evolution of Knowledge: Beyond the Biological Model', Research Policy 31: 1227–39.

Loeb, J. (1897–1898) 'Einige Bemerkungen über den Begriff, die Geschichte und Literatur der allgemeinen Physiologie', Pflügers Archiv für die gesamte Physiologie 69: 249–67.

Luhmann, N. (1998) Die Gesellschaft der Gesellschaft (Frankfurt: Suhrkamp).

Martin, W.F. (2011) 'Early Evolution without a Tree of Life', Biology Direct 6: 36 (http://www.biology-direct.com/content/6/1/36) [last access 08.05.2013].

Mauelshagen, F. (2003) 'Networks of Trust: Scholarly Correspondence and Scientific Exchange in Early Modern Europe', The Medieval History Journal 6: 1–32.

Merton, R.K. (1973) The Sociology of Science: Theoretical and Empirical Investigations (Chicago: University of Chicago Press).

Minsky, M. (1985) The Society of Mind (New York: Simon & Schuster).

Minsky, M. (1990) Mentopolis (Stuttgart: Klett-Cotta).

Mücke, M. & T. Schnalke (2009) Briefnetz Leopoldina: die Korrespondenz der Deutschen Akademie der Naturforscher um 1750 (Berlin/New York: Walter de Gruyter).

Murray, F. (2002) 'Innovation as Co-evolution of Scientific and Technological Networks: Exploring Tissue Engineering', Research Policy 31: 1389–403.

Nelson, C. (1997) 'Superstars', Academe 83(1): 38–43, 54.

Newman, M.E.J. (2010) Networks: An Introduction (Oxford/New York: Oxford University Press).

Oeser, E. (1988) Das Abenteuer der kollektiven Vernunft: Evolution und Involution der Wissenschaft (Berlin: P. Parey).

Osterhout, W.J.V. (1928) 'Jacques Loeb', The Journal of General Physiology 8: ix-lix.

Pauly, P.J. (1987) Controlling life: Jacques Loeb and the Engineering Ideal in Biology (New York: Oxford University Press).

Pearl, J.L. (1984) 'The Role of Personal Correspondence in the Exchange of Scientific Information in Early Modern France', Renaissance and Reformation 20: 106–13.

Plotkin, H.C. (1982) Learning, Development, and Culture: Essays in Evolutionary Epistemology (Chichester West Sussex, England/New York: J. Wiley).

Popper, K.R. (1979) Objective Knowledge: an Evolutionary Approach (Oxford: Oxford University Press).

Popper, K.R., K. Lorenz & F. Kreuzer (1985) Die Zukunft ist offen: das Altenberger Gespräch (München: Piper).

Radnitzky, G. & W. Warren Bartley (1987) Evolutionary Epistemology, Rationality, and the Sociology of Knowledge (La Salle Ill.: Open Court).

Reitmeyer, M. & C. Marx (2010) 'Netzwerkansätze in der Geschichtswissenschaft', in C. Stegbauer & R. Häußling (eds), Handbuch Netzwerkforschung (Wiesbaden: VS Verlag für Sozialwissenschaften): 869–80.

Richards, R.J. (1987) Darwin and the Emergence of Evolutionary Theories of Mind and Behavior (Chicago: University of Chicago Press).

Riedl, R. & R. Kaspar (1980) Biologie der Erkenntnis: die stammesgeschichtlichen Grundlagen der Vernunft (Berlin/Hamburg: Parey).

Rosch, E.H., C.B. Mervis, W.D. Gray, D.M. Johnson & P. Boyes-Braem (1976) 'Basic objects in natural categories', Cognitive Psychology 8: 382–439.

Rusnock, A. (1999) 'Correspondence Networks and the Royal Society, 1700–1750', British Journal for the History of Science 32(113): 155–69.

Steinke, H. (2004) 'Why, what and how? Editing early modern scientific letters in the 21st century', Gesnerus 61: 282–95.

Steinke, H. & M. Stuber (2004) 'Medical Correspondence in Early Modern Europe. An Introduction', Gesnerus 61: 139–60.

Sterelny, K. (1994) 'Science and Selection', Biology and Philosophy 9: 45–62.

Stock, W. (1985) 'Die Bedeutung der Zitatenanalyse für die Wissenschaftsforschung', Zeitschrift für allgemeine Wissenschaftstheorie 16(2): 304–14.

Strauß, G. (1996) 'Wort – Bedeutung – Begriff: Relationen und ihre Geschichte', in J. Grabowski, G. Harras & T. Herrmann (eds), Bedeutung – Konzepte – Bedeutungskonzepte. Theorie und Anwendung in Linguistik und Psychologie (Opladen): 22–46.

Thagard, P. (1980) 'Against Evolutionary Epistemology', PSA: Proceedings of the Biennial Meeting of the Philosophy of Science Association 1:187–96.

Toulmin, S. (1972) Human Understanding (Princeton, N.J.: Princeton University Press).

Twain, M. (1923) 'Dr. Loeb's Incredible Discovery', in M. Twain (ed), The Complete Works of Mark Twain. Europe and Elsewhere (New York: Harper and Brothers): 304–09.

Vaihinger, H. (1924) The Philosophy of 'as if', a System of the Theoretical, Practical and Religious Fictions of Mankind (London/New York: K. Paul, Trench, Trubner & Co., Harcourt, Brace & Company, inc.).

Verworn, M. (1895) Allgemeine Physiologie: ein Grundriss der Lehre vom Menschen (Jena: Fischer).

Vollmer, G. (1975) Evolutionäre Erkenntnistheorie: angeborene Erkenntnisstrukturen im Kontext von Biologie, Psychologie, Linguistik, Philosophie u. Wissenschaftstheorie (Stuttgart: Hirzel).

Vollmer, G. (1987) 'What Evolutionary Epistemology is not', Synthese Library 190: 203–21.

Wetherell, C. (1998) 'Historical Social Network Analysis', International Review of Social History 43: 125–44.

White, H.C. (2008) Identity and Control: How Social Formations Emerge (Princeton: Princeton University Press).

White, H.D. (2000) 'Toward Ego-Centered Citation Analysis', in B. Cronin & H.B. Atkins (eds), The web of knowledge: a festschrift in honor of Eugene Garfield (Medford, NJ: Information Today): 475–96.

White, H.D. (2001) 'Authors as citers over time', Journal of the American Society for Information Science and Technology 52(2): 87–108.

Whitley, R. (1984) The Intellectual and Social Organization of the Sciences (Oxford: Clarendon Press).

Wuketits, F.M. (1983) Concepts and Approaches in Evolutionary Epistemology: Towards an Evolutionary Theory of Knowledge (Dordrecht/Boston: Reidel / Kluwer Academic Publishers).

Zimmer, D.E. (1980) 'Ich bin, also denke ich. Die biologischen Wissenschaften haben sich in die Philosophie eingemischt', Die Zeit 24, 06.06.1980.

BRIDGING DISCIPLINES.
CONCEPTUAL DEVELOPMENT IN INTERDISCIPLINARY GROUPS

Hanne Andersen

The current volume on classification and evolution in biology, linguistics and history of science is an example of the kind of interdisciplinary collaboration that we see in science today and in which multiple scientists with different areas of expertise share and integrate their cognitive resources in producing new results that cut across disciplinary boundaries. But how *do* scientists involved in interdisciplinary collaborations link concepts originating in different disciplines or research fields, and how *do* they develop new concepts that cut across disciplinary boundaries? How do biologists, linguists and historians of science working on evolution and classification in their various fields come to collaborate on network models as seen in this volume, and how do they create links between their concepts of species and languages? While the scientific details on these developments are found in the preceding chapters in this volume, I shall in this chapter provide some general reflections on how scientists involved in interdisciplinary collaboration link concepts originating in different disciplines or research fields.

   I shall starts from the philosopher of science Thomas Kuhn's account of conceptual development and the later re-interpretation in terms of dynamic frames that has been developed by Peter Barker, Xiang Chen and me (Andersen et al. 2006). Drawing on Kuhn's ideas about linguistic disparity between different scientific communities, much work on interdisciplinary collaboration has been based on the assumption that the relation between different disciplines can be characterized by incommensurability and that, therefore, there is a deep problem of communication when entering interdisciplinary collaboration. However, I shall argue that the conceptual disparity between different disciplines or specialties is different from the conceptual disparity that are usually describe by incommensurability, and that the overly strong claim about conceptual disparity ignores exactly the kind of conceptual developments across disciplines that is witnessed in collaborations like the one witnessed in the current volume. I shall therefore show how Kuhn's account needs to be extended to include the many relations that exist *between* lexicons. Based on this extended account I shall show how scientists collaborating in interdisciplinary research combine their conceptual resources and adopt structures and constraints provided by collaborators from other areas of expertise. Finally, I shall use this extended Kuhnian model of conceptual structures to give a re-interpretation of the currently popular notions of trading zones and boundary objects and indicate some of the questions that can be raised which analyzing interdisciplinary collaborations like the one displayed in this volume.

KUHN'S ACCOUNT OF CONCEPTUAL STRUCTURES IN SCIENCE

Throughout his academic career, Thomas Kuhn worked on developing an account of scientific concepts that supported and substantiated his ideas about the development of science as phases of normal sciences interrupted by scientific revolutions. In his work on scientific concepts, Kuhn focused almost exclusively on taxonomic terms, that is, terms "which refer to the objects and situations into which a language takes the world to be divided" (Kuhn 1990: 4). On this view, a taxonomic conceptual structure, or a lexicon as Kuhn also called it, is "a more general sort of categorizing module" (Kuhn 1990: 5) in which 'certain sorts of expectation about the world are embedded' (cf. Kuhn 1990: 8). Thus, a lexicon divides objects into groups according to their similarity and dissimilarity, and if represented in a tree structure with concepts as the branch points, a set of features useful for distinguishing among objects at the next level down is attached to each branch point or node.

Kuhn's account of concepts is basically a family resemblance account according to which conceptual structures build on relations of similarity and dissimilarity between perceived objects. On this account, special importance is ascribed to *dissimilarity* and the properties which differentiate between instances of *contrasting* concepts, that is, concepts whose instances are more similar to one another than to instances of other concepts and which can therefore be mistaken for each other (see e.g. Kuhn 1979: 413). Because the instances of contrasting concepts are more similar to one another than to instances of other concepts, the set of contrasting concepts also form a family resemblance class at the super ordinate level, and family resemblance concepts therefore form hierarchical structures in which a general concept decomposes into more specific concepts that may again decompose into yet more specific concepts, and so forth – in other words, taxonomies.

The expectations about the world that are embedded in a lexicon are assumptions about what exist and does not exist (what Hoyningen-Huene (1992) has termed the quasi-ontological knowledge) and assumptions about the empirical correlations of features (what Hoyningen-Huene has termed knowledge of regularities). Further, as Andersen et al. (2006) have argued, drawing on the so-called theory-theory developed within cognitive science, these correlations of features are not simply empirical generalizations, instead, theories will usually be developed that explain the correlations, and further, those features that play a causal role in explaining other features tend to be seen as more important.

Drawing on the work of the cognitive psychologist Barsalou, Andersen, Barker and Chen have argued that in the analysis of taxonomies the various features are not all equal (Andersen et al. 1996; Chen et al. 1998; Barker et al. 2003). Instead, some features are values of others, and it is differences between such different values of the same attribute that distinguish between contrasting categories. An adequate representation of this kind of taxonomy is Barsarlou's dynamic frame representation (Barsarlou 1992) that emphasizes the distinction between values and attributes and represents empirical correlations as value constraints.

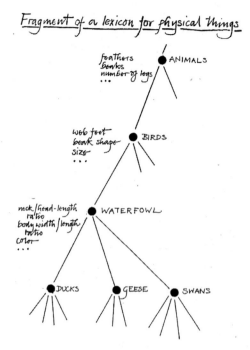

*Figure 1. Kuhnian taxonomy of waterfowl as a subordinate to birds
and super ordinate to ducks, geese, and swans (Kuhn 1990).*

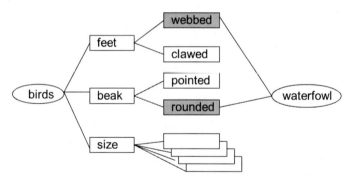

Figure 2. Dynamic frame for waterfowl as a subordinate to the concept bird.

Kuhn's own examples were primarily simple concepts like the contrast set of waterfowl that may appear different from scientific concepts like the physical concept
"force" or the biological concept "evolution". In his later writing Kuhn introduced
a distinction between *normic* concepts that are "learned as members of one or another contrast set" and *nomic* concepts that "stand alone" (cf. Kuhn 1993: 317).
When emphasizing that the ability to pick out referents of normic terms "depends
critically upon the characteristics that differentiate its referents from those of the

other terms in the set, which is why the terms involved must be learned together and why they collectively constitute a contrast set" (Kuhn 1993: 317) while nomic terms "are learned from situations in which they occur together exemplifying laws of nature" (ibid.) it might seem as if Kuhn suddenly saw a need to introduce a distinction between similarity class concepts and non-similarity class concepts. However, as argued by Andersen and Nersessian (2000), both normic and nomic concepts can be understood as similarity class concepts, but in the case of nomic concepts the family resemblances are among *complex problem situations* rather than among *individual objects* or phenomena. Nersessian (1984) has argued that nomic concepts can be represented by a "meaning schema" which is a frame-like structure in which a scientific concept is represented by four components central to its descriptive and explanatory function: ontological status, function, mathematical structure, and causal power. Andersen and Nersessian (2000) have shown how this meaning schema can be linked to the frame representation of concepts. The causal power of a concept marks out the problem situations in which the concept comes into use in order to explain the situation. Hence, this component of the meaning schema can be linked to the frame representing a similarity class of problem situations. Likewise, the mathematical structure corresponds to the scientific laws associated with this similarity class of problem situations. The additional components of the meaning schema, 'function' and 'ontological status', are the components which serve to distinguish individual concepts within the complex situation. The function of a concept marks out a specific part of the explanation of a problem situation and clarifies its explanatory role. To the various functions corresponds an ontological status, that is, a belief about what kind of 'stuff' is responsible for this particular function. For example, on this view of nomic concepts one can reconstruct the similarity class of problem situations in which hypotheses about historical patterns of descent are constructed and evaluated in the form of evolutionary trees. In these problem situations, several concepts will be distinguished by their function and ontological status, among them ancestry, descent, and evolutionary relatedness or root, branch, and internal and terminal nodes.

CONCEPTS AND COMMUNITIES

So far, this account of concepts has said little about the relation between conceptual structures and the cognizing scientists who hold these concepts. However, ever since *The Structure of Scientific Revolutions* Thomas Kuhn had been interested in the relation between a scientific community and the conceptual structures that the members of this community hold. In the Postscript to the second edition of *Structure*, Kuhn saw a circularity in the way he had introduced scientific communities and the various cognitive elements that he at that point still referred to with the overarching concept of paradigms, namely that "a paradigm is what the members of a scientific community share, and conversely a scientific community consists of men who share a paradigm" (Kuhn 1970: 176). Later, after Kuhn had replaced the diffuse notion of paradigms with the more specific notion of lexicons as the description of the cognitive resources that the members of a scientific community share, he argued that what the members of a scientific community share is the *general structure* of the scientific lexicon:

> "[...] a community of intercommunicating specialists, a unit whose members share a lexicon that provides the basis for both the conduct and the evaluation of their research and which simultaneously, by barring full communication with those outside the group, maintains their isolation from practitioners of other specialties." (Kuhn 1991: 8)

Hence, on Kuhn's view, members of a scientific community share the overall structure of their lexicon although the individual details may differ from scientist to scientist. If, on the contrary, the overall structure differs, they will have different expectations about the world and communication difficulties will arise:

> "What members of a language community share is homology of lexical structure. Their criteria need not be the same, for those they can learn from each other as needed. But their taxonomic structures must match, for where structure is different, the world is different, language is private, and communication ceases until one party acquires the language of the other" (Kuhn 1983: 683)

Later, this led Kuhn to the idea that during the historical development of science new subspecialties emerge and gradually get isolated from each other due to a growing conceptual disparity between the developed tools and he claimed that this incommensurability was what keeps scientific disciplines apart (Kuhn 1990). However, as I have argued elsewhere (Andersen 2006), this new role ascribed to incommensurability revives the incommensurability problem as it was originally raised by Shapere (1971), namely how to make sense of the idea that incommensurable theories are actually competing. The conceptual disparity between two specialties placed at different branches of the evolutionary tree of the sciences is different in important ways from the conceptual disparity between the two specialties at each side of a revolutionary divide. Lack of communication between different specialties does not necessarily reflect incommensurability, but may simply reflect the fact that the two address issues that are, at least so far, unconnected. In addition, neither does this account include cases in which different disciplines or subspecialties populated by disjunct communities happen to have developed homologous lexical structures, despite addressing different domains, like, for example, numerical phylogenetics and lexicostatistics in the 1950es and 1960es that seem to have developed independently, although the parallels were also noted by the historical actors at the time (Atkinson and Gray 2005). However, these shortcomings of Kuhn's account open for a new set of challenging and interesting questions relating to communities and conceptual structures, namely how conceptual structures originating in different disciplines can be brought to connect in the course of interdisciplinary collaboration, and how these connections can be distributed in the relevant communities.

CONCEPTUAL STRUCTURES ACROSS DISCIPLINES

Kuhn's view about conceptual disparity between distinct specialties reflects the classical view of unidisciplinary competence, that is, the image of scholars being fully competent in one discipline and competent in this discipline only. However, as argued by, among others, Campbell (1969), a discipline consists of a congerie of narrow specialties, and the integration of these specialties into a comprehensive discipline is a *collective* product of the discipline's community and not something embodied within the individual practitioner. It is achieved through the fact that the multiple narrow specialties overlap and that through this overlap, a collective communication, a collective competence and breadth, is achieved. In addition, the claim about conceptual disparity between disciplines ignores the kind of conceptual developments which involve non-competing but in various ways related specialties.

In order to analyze the process through which scientists with overlapping conceptual structures collaborate and integrate knowledge drawn from their respective areas of expertise, we need to extend Kuhn's account in several ways. First, the Kuhnian account of concepts needs to be extended in the way that the same concept may form part of multiple taxonomies based on different relations of similarity and difference. Second, taxonomies may be interconnected in complicated criss-crossing patterns because concepts are related to other concepts imbedded in different lexicons, for example through regularities or laws. Third, features used to distinguish between nodes in one taxonomy will often themselves be nodes in other taxonomies. These various forms of conceptual relations across scientific lexicons can reveal important details about how scientists can collaborate across fields beyond what is offered by currently popular accounts.

BOUNDARY OBJECTS AND TRADING ZONES

In describing how scientists from different fields collaborate, two kinds of descriptions have become popular: descriptions in terms of *trading zones*; a notion introduced by the historial of science Galison in his 1997 monograph *Image and Logic* and descriptions in terms of *boundary objects*; a notion introduced by the sociologists of science Star and Griesemer in a paper in *Social Studies of Science* in 1989.

Galison introduced the idea of trading zones based on his studies of modern physics where he saw different subcultures of instrumentation, experiment and theory; subcultures that shared some activities while diverging on many others. The important point was that despite differences in classification among these subcultures, they could collaborate within a local context which Galison termed a trading zone and which he saw as "a social, material, and intellectual mortar binding together the disunified traditions of experimenting, theorizing, and instrument building" (Galison 1997: 803). Such a trading zone need not be permanent; instead it may be productive for a while and then die out again. Further, the different collaborating groups may maintain their distinct character while coordinating their approached around specific practices (cf. Galison 1997: 806).

An important aspect of the trading zone is that the collaborating groups all impose constraints on the nature of the exchange. Galison's aim was to argue against incommensurability by showing that although concepts may change over a scientific revolution, such as the change in the concept of mass from Newtonian mechanics to Einsteinian relativity theory, there is still a localized zone of activity in which experimenters and theorists could coordinate beliefs and actions. Or, in terms of another of Galison's examples, the research on the recombination of quarks into jets of observable particles:

> "we have accounts of phenomena in which terms (jets, quarks, partons, gluons, hadronization) are used in such heterogeneous ways and on the face of it carry such different meanings that we might expect to locate them in different and incommensurable conceptual schemes or paradigms. And yet, once again, we have a site at which the actors worked furiously to coordinate and adjudicate among alternatives" (Galison 1997: 814)

This coordination of action occurs by use of a contact language constructed with the elements of the languages of the collaborators as a pidgin language in which parts of the fuller languages are withheld.

> "Reduction of mathematical structure, suppression of exceptional cases, minimization of internal links between theoretical structures, simplified explanatory structure – these are all ways that the theorists prepare their subjects for the exchange with their experimental colleagues" (Galison 1997: 835).

Drawing on the extended Kuhnian account of conceptual structures, we may understand points of contact between different lexicons in the way that only some differentiating features are emphasized in the communication across disciplinary boundaries, namely such features that can easily be recognized by the members of the other community. However, that does not imply that other features are discarded. On the contrary, the members of each community do, as Galison notes, 'work furiously to coordinate and adjudicate'. Although they may reduce and simplify mathematical and explanatory structures in the direct communication across the disciplinary boundary, each discipline will be focused on how to encompass the results that are exchanged *between* disciplines into the much richer structure that is shared *within* each discipline.

Similarly, Star and Griesemer's (1989) introduced the notion of boundary objects to denote concepts which are plastic enough to adapt to local needs and to the constraints of the actors employing them, but at the same time also robust enough to maintain a common identity across sites.

> "Boundary objects are objects which are both plastic enough to adapt to local needs and the constraints of the several parties employing them, yet robust enough to maintain a common identity across sites. They are weakly structured in common use and become strongly structured in individual use. These objects may be abstract or concrete. They have different meanings in different social worlds, but their structure is common enough to more than one world to make them recognizable, a means of translation" (Star and Griesemer 1989: 393).

Describing how much scientific work on the one hand is conducted by diverse groups of actors, on the other hand requires cooperation between these diverse actors, Star and Griesemer has argued that there is a central tension in science between divergent viewpoints and the need for generalizable findings (cf. Star and

Griesemer 1989: 347), and with their notion of boundary objects they emphasize locality, plasticity and simplification as important characteristics of communication across disciplines.

These two notions of boundary objects and trading zones have been combined in recent work by Collins, Evans and Gorman (Collins et al. 2006; Gorman 2010) who have argued that interdisciplinary collaboration can be analyzed as trading zones varying along two dimension: a cultural dimension according to the degree of linguistic homogeneity or heterogeneity, and a power dimension according to the degree to which power is used to enforce the collaboration. On their account, some trading zones, what they call *inter-language trading zones*, result in a truly merged culture in which a full blown creole language is the ideal end process. Other forms of trading zones are *enforced trading zones* in which the expertise of an elite group is black-boxed for other participants, *subversive trading zones* where one language overwhelms that of the other, and *fractionated trading zones* which may as a trading zone be mediated either by material culture in the form of boundary objects, or by language in the form of interactional expertise. Interactional expertise is here a notion developed by Collins to denote the level of expertise sufficient to interact in interesting ways with participants of another specialty or discipline but without having the contributory expertise required to contribute to the research of this field (see e.g. Evans and Collins 2010).[1] This is acquired gradually through linguistic socialization in which one party learns the scientific language of the other while retaining their own distinct contributory expertise. Similarly, in the case of boundary objects, the collaborating disciplines or specialties remain distinct and each work with the object(s) in their own way and impose their own meaning to it.

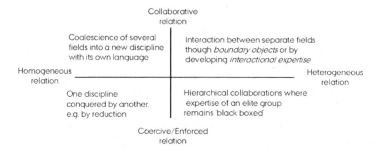

Figure 3. Four types of trading zones as described by Collins, Evans & Gorman (2006).

Importantly, these four kinds of trading zones are not static forms. For example, interdisciplinary work may start with a fractionated trading zone (mediated either by boundary objects or by interactional expertise), but as collaboration intensifies the cultural differences between collaborators from different disciplines may be reduced and the fractionated trading zone gradually transform into an inter-lan-

---

1   The notion was originally developed to describe Collins experience as a sociologists studying physicists and gradually learning to interact with them to such a degree that he could, for example, conduct conversations with them on their research, but without himself being able to perform experiments etc.

guage trading zone in which a full-blown creole language is the ideal end point. As an example, Collins, Evans and Gorman (2006) explain how an imaginary research group may start from a slightly enforced encouragement from the outside research system to develop a collaborative research application. In the initial steps of the collaboration, the application itself works as a boundary object that may mean different things to the various collaborators, although these differences will not be so important that they undermine the joint project. As the work intensifies, it gradually becomes increasingly voluntary and moves upwards in the diagram. Further, as collaboration increases, the collaborators may become so interested in each other's' work that they want to understand more about it and they start developing interactional expertise. Later, they may begin to invent jargon terms, and the disciplinary difference between the different participants will be reduced. As their work become more and more homogeneous, the trading zone moves to the left in the diagram and gradually develops into an interlanguage trading zone and may eventually develop into a distinct, new discipline.

The interdisciplinary collaboration between biologists, linguists and historians of science that lies behind this volume seems to reflect several of the patterns described above, although it also differs in important ways. Obviously, it is not possible to give a detailed account of this full history, and some suggestive remarks and questions will have to suffice as the conclusion of this paper. [2] First of all, contrary to the situation described by Collins, Evans and Gorman, the current collaboration is not establishing a completely new trading zone. There is a long history of parallel developments as well as mutual influence between evolutionary biology and historical linguistics, as described in previous chapters in this volume. [3] A full analysis of the contemporary trading zone would therefore have to build also an account of previous developments of a pidgin or creole language for the description of evolutionary trees, including what kinds of simplifications or minimization of internal links have been made in communication with practitioners from the other discipline, and where the shared communication language has remained a pidgin language where details were unfolded only within each discipline, or developed into a creole where the two disciplines have mutually influenced each other's conceptual development.

Second, an important element in the current trading zone is the recent focus on network structures that in contrast to tree structures can accommodate horizontal interactions. This form of network reconstruction seems to function as a boundary concept that, after being developed in evolutionary biology to take horizontal transfer of genes during microbial evolution into account, is now adapted to the needs of historical linguistics to take lexical borrowing during linguistic evolution into account. This work draws on highly technical methods that would make one expect elements of the collaboration to resemble Galison's description of initial communication in a pidgin language with reduced mathematical structures, minimization of

2   I would like to thank the members of the EvoClass collaboration for giving me access to project material.
3   See also Sommerfeld and Kressing 2011.

internal links, simplified explanatory structures, etc. – but while retaining the full structure and complexity within the originating discipline.

That leads immediately to a third, important aspect, namely the direction of the trading zone's development where the scheme developed by Collins, Evans and Gorman may point to important reflections on the future direction of the collaboration. In how far are collaborators gradually acquiring interactional expertise through a linguistic socialization in which one party learns the language of the other while at the same time retaining their contributory expertise in their original field? In how far is the current collaboration between evolutionary biology and historical linguistics striving for the development of an inter-language trading zone with a truly merged culture and a full blown creole language? Would such a full blown creole be that of a collaborative relation between equals with elements drawn from the languages of all participating disciplines, or would it be a coercive relation in which the language of one discipline comes to dominate that of the others? The many recollections during the meeting in which this volume originates about the struggle to learn each other's scientific language at the previous meetings suggest that this development of interactional expertise has played an important role in the collaboration, and interesting lessons for interdisciplinary collaborations between the sciences and the humanities may be drawn from these experiences.

REFERENCES

Andersen, H. (2006) 'How to recognize introducers in your niche', in H.B.Andersen, F.V. Christiansen, K.V. Jørgensen & V. Hendricks (eds), The Way Through Science and Philosophy: Essays in Honour of Stig Andur Pedersen (London: College Publications): 119–36.

Andersen, H., Barker, P. & X. Chen (1996) 'Kuhn's mature philosophy of science and cognitive science', Philosophical Psychology 9: 347–63.

Andersen, H., Barker, P. & X. Chen (2006) The Cognitive Structure of Scientific Revolutions (Cambridge, MA: University Press).

Andersen, H. & N.J. Nersessian (2000) 'Nomic concepts, frames and conceptual change', Philosophy of Science (Proceedings) 67: 224–41.

Atkinson, Q.D. & R.D. Gray (2005) 'Curious Parallels and Curious Connections – Phylogenetic Thinking in Biology and Historical Linguistics', Syst.Biol. 54: 513–26.

Barker, P., Chen, X. & H. Andersen (2003) 'Thomas Kuhn and Cognitive Science.' in T. Nickles (ed), Thomas Kuhn (Cambridge, MA: University Press): 212–45.

Barsalou, L.W. (1992) Cognitive Psychology: An Overview for Cognitive Scientists. (Hillsdale, NJ: Erlbaum).

Barsalou, L.W. (1992) 'Frames, concepts, and conceptual fields', in A. Lehrer & E.F. Kittay (eds), Frames, Fields, and Contrasts (Hillsdale, NJ: Erlbaum): 21–74.

Campbell, D.T. (1969) 'Ethnocentrism of disciplines and the fish-scale model of omniscience', in M. Sherif & C.W. Sherif (eds), Interdisciplinary relationships in the social sciences (Chicago: Aldine): 328–48.

Chen, X., H. Andersen & P. Barker (1998) 'Kuhn's Theory of Scientific Revolutions and Cognitive Psychology', Philosophical Psychology 11: 5–28.

Collins, H., R. Evans & M. Gorman (2006) 'Trading zones and interactional expertise', Studies in History and Philosophy of Science 38: 657–66.

Evans, R. & H. Collins (2010) 'Interactional Expertise and the Imitation Game' in M.E. Gorman (ed), Trading Zones and Interactional Expertise. Creating New Kinds of Collaboration (Cambridge, MA: MIT Press): 53–70.

Galison, P. (1997) Image and Logic. A Material Culture of Microphysics (Chicago: University Press).

Gorman, M. E. (2010) Trading Zones and Interactional Expertise (Cambridge, MA: MIT Press).

Hoyningen-Huene, P. (1992) Reconstructing Scientific Revolutions (Chicago: University Press).

Kuhn, T. S. (1970) The Structure of Scientific Revolutions (Chicago: University Press).

Kuhn, T. S. (1979) 'Metaphor in Science' in A. Ortony (ed), Metaphor in Science (Cambridge, MA: University Press): 409–19.

Kuhn, T. S. (1983) 'Commensurability, Comparability, Communicability', PSA 2: 669–88.

Kuhn, T. S. (1990) An Historian's Theory of Meaning, Ref Type: Unpublished Work.

Kuhn, T. S. (1990) 'The Road Since Structure', PSA 2: 3–13.

Kuhn, T. S. (1991) 'The Road since Structure,' PSA 2: 3–13.

Kuhn, T. S. (1993) 'Afterwords', in P. Horwich (ed), World Changes (Cambridge, MA: MIT): 311–41.

Nersessian, N. J. (1984) Faraday to Einstein: Constructing Meaning in Scientific Theories (Dordrecht: Martinus Nijhoff).

Shapere, D. (1971) 'The paradigm concept', Science 172: 706–9.

Sommerfeld, S. & F. Kressing (2011) Die Sprachen-Spezies. Analogie in Biologie und Linguistik, Working Paper.

Star, S. L. & J. R. Griesemer (1989) 'Institutional Ecology, 'Translations' and Boundary Objects: Amateurs and Professionals in Berkeley's Museum of Vertebrate Zoology, 1907–39', Social Studies of Science 19: 387–420.

HISTORICAL NETWORK ANALYSIS CAN BE USED TO CONSTRUCT A SOCIAL NETWORK OF 19TH-CENTURY EVOLUTIONISTS[1]

Matthis Krischel, Heiner Fangerau

1. INTRODUCTION

In the village Down, an hour outside of London, one can visit a recreation of Charles Darwin's personal study on the ground floor of his former home (figure 1). Indicators of intellectual and social exchange can be found in the study: a bookcase reflected in the mirror above the fireplace contains books and, perhaps, bound journals published by learned societies. A stack of letters is on the desk, and vessels on the round table may contain plant or animal specimens, some of which might have been sent from halfway around the world. One can imagine Darwin sitting in the comfortable chair and receiving guests.

With this image in mind, we want to show in this contribution, how historians can use social network analysis to quantify and visualise the intellectual and social exchange that was part of daily scientific practice for 19th century evolutionists, as it is for scientists today. To a degree, this approach breaks down the barrier between hermeneutic analysis, which is common in the humanities, and computer-aided quantitative analysis, which is more common in the social and natural sciences.

Communication and the exchange of ideas have been central features of science since at least the 17th century.[2] Scientists' interactions with one another and the outside world can assume two forms that often go hand in hand: through the scientific literature and through personal contacts, such as by correspondence (Rusnock 1990: 155–69; Pearl 1984: 106–13; Steinke and Stuber 2004: 139–60; Mauelshagen 2003: 1–32). The attempt to reconstruct scientists' personal and intellectual contacts yields a topography of actors constituting a network that goes far beyond simple personal relationships. First, members of the network are linked not only to one another directly, but also indirectly through other highly influential members. Second, because each individual in the network represents a research identity, the network can be understood as a scientific web, with clusters of people representing

1   This contribution is based in part on a paper presented at the Bridging Disciplines Conference, held 24–26 June 2011 at Wissenschaftszentrum Schloss Reisensburg, Germany; a poster presented at the History of Science Society meeting, held 3–6 November 2011 in Cleveland, OH, USA; and a paper presented at the Falling Walls Lab, held 8 November 2012 in Berlin, Germany. The authors would like to thank everyone who provided feedback on those occasions.
2   An example is knowledge exchange in the so-called "republic of letters". For an overview, see (among others): Casanova 2004 and Dalton 2003.

scientific communities, disciplinary or interdisciplinary knowledge (cf. Fangerau, this volume; Fangerau 2009).

Previously it has been pointed out that the theory of descent with modification, as formulated in the 19th century, owes much to the interdisciplinary nature of research in that era. Natural philosophers and philologists exchanged ideas with sociologists, anthropologists, and economists, generating new ideas (Kressing et al. 2011). Modern evolutionary researchers have largely forgotten this heritage, leading to the re-emergence of some of the potential problems of classifying human biodiversity and cultures using similar methods – particularly the assumption that the two are linked. In the second half of the 20th century, quantitative linguists and population geneticists again began to borrow approaches from one another and collaborate on research projects (Krischel et al. 2013).

In this contribution, we reconstruct the social network of evolutionary theorists centred around Darwin, and further describe the larger network of 19th-century evolutionists comprised of linguists, biologists, anthropologists, and other scholars. The term "network" is used not only as a metaphor for contact among historical actors, but to refer to a social network reconstructed on the basis of historical sources that is described, visualised, and analysed below.

Figure 1: Recreation of Charles Darwin's study at Down House,
image courtesy of English Heritage

2. EVOLUTION AND CLASSIFICATION IN 19TH-CENTURY LINGUISTICS AND BIOLOGY

Studies of languages and living forms share some problems. Scholars in both disciplines are presented with large numbers of study objects, some of which seem remarkably similar, while others are obviously very different. Natural historians and linguists have sought to impose order on the perceived diversity by establishing classification systems and explaining development. In biology, the history of classification has received much less attention than that of evolutionary theories (for an overview of the history of classification in biology, cf. Farber 2000; for an overview of the history of evolutionary theory, cf. Bowler 2003; Larson 2006). Historians have pointed out intellectual and personal connections between natural historians (i.e. scholars concerned with biological evolution) and comparative linguists in the 19th century (cf. Alter 2002), as well as those between linguists and biological anthropologists and the ways in which their work was used to argue for the inequality of human populations (cf. Römer 1985).

The basic model of evolution, meaning slow, gradual change from simpler to more complex forms, has roots reaching back at least to the Enlightenment. According to Kenneth Bock, it represents "a mode of conceiving change that is deeply rooted in Western thought" (Bock 1955: 133). He stated: "The classical view of change as growth, the seventeenth century idea of progress, eighteenth century conjectural or hypothetical histories, and nineteenth century evolutionism all share in the perspective that change is natural, inevitable, slow, gradual and continuous." (Bock 1955: 129) Good examples of Bock's exposition can be found in the works of several scholars from Scotland and France, where the "comparative study of societies and how change in general takes place" was undertaken during the Enlightenment (Trigger 1998: 32), including Adam Ferguson's *Essay on the history of civil society* (1767) and Marie-Jean Antoine de Condorcet's *Esquisse d'un tableau historique des progrès de l'esprit humain* (1795).

In the 19th century, evolutionary theories proliferated in the developing discipline of biology. Jean-Baptiste de Lamarck (1809), Robert Chambers (anonymously, 1844) (cf. Secord 2000), and Darwin (1859) published important contributions within five decades. In the *Origin of Species*, Darwin offered a solution to the problems of explaining biological development and classifying diversity. In chapter 13, under the sub-heading "Classification", he wrote: "Thus, on the view which I hold, the natural system is genealogical in its arrangement, like a pedigree" (Darwin 1859: 422). He later elaborated: "I believe this element of descent is the hidden bond of connexion which naturalists have sought under the term of the Natural system. [...] We can clearly see how it is that all living and extinct forms can be grouped together in one great system" (Darwin 1859: 433). By establishing the pedigree as an answer to the question of the origin of species – several modern species are descended from one older species, and, in turn, several older species are descended from an even older one – as well as the question of classification – it should be based on common descent at the various levels of the pedigree –, Darwin demonstrated the wide applicability of his theory and by using it as the only illustration in the *Origin of Species*, engrained the tree of life as an exemplary symbol.[3]

Common descent had also become the established paradigm in comparative linguistics by the early 19th century. The development of polyphyletic models of language origin, such as Marcus von Boxhorn's first identification of the language family now referred to as "Indo-European", began in the 17th century. In 1647, Boxhorn formulated a theory of language families based on extinct ancestral languages (cf. Kressing et al. 2013). In 1786, William Jones described the common root of Sanskrit, Greek, and Latin, which is often considered to be the starting point

---

3   "Exemplar" is here used in the sense of Kuhn (1970).

of comparative studies in Indo-European languages (cf. Lockwood 1969: 22). Historical comparative linguistics became more established in the 19th century, resulting in the reconstruction of proto-languages according to laws of regular sound correspondences (Krischel et al. 2011: 110).

Scholars began to use trees to illustrate classifications of species and languages around 1800. Some historians have recently examined the use of pedigrees as models for such classifications (for an example from the history of biology, cf. Archibald 2009.) An early tree-like diagram in biology illustrated the relationships among living forms in Augustin Augier's *Essay d'une nouvelle classification des vegetaux* (1801), in which he stated:

> "A figure like a genealogical tree appears to be the most proper to grasp the order and gradua-
> tion of the series of branches which form classes or families. This figure, which I call a botani-
> cal tree, shows the agreements which the different series of plants maintain among each other,
> although detaching themselves from the trunk, just as a genealogical tree shows the order in
> which different branches of the same family came from the stem to which they owe their
> origin."[4]

Similarly, an engraving depicting an *Arbre généalogique des Langues mortes et vivantes* was made for the Abbé Sicard, a French linguist, around 1800 (Mauro et al. 1990).

Contacts between linguists and biologists in the second half of the 19th century have been well documented (cf. Koerner 1981; Taub 1993; Richards 2002; Alter 2002; Krischel et al. 2011). Stephen Alter argued that Darwin's approach not only fit the 19th-century *Zeitgeist*, insofar as his historically minded study of the origin and relatedness of species fit into the "antiquarian ethos [which] united much of the era's scholarship, transcending boundaries between the sciences and the humanities", but also that his recourse to "philology and its allied disciplines helped construct the scaffolding of plausibility surrounding the house of Darwin" (Alter 2002: 148). Alter identified the linguistic metaphor as "seemingly natural" for the mid-19th century. He argued that the discipline of comparative philology was in its heyday at that time, which meant that a large educated public was familiar with notions of gradual change over time and "the slow transformation of languages provided an apt analogy for the gradual transmutation of species" (Alter 2002: 2). At the same time, the notion of a common ancestral language for at least large language families, such as Indo-European, was commonly accepted in comparative linguistics. By employing the analogy used in linguistics, where descent with modification was established, Darwin hoped to transfer this credibility to his theory of descent with modification of plants and animals. Alter mentioned that Darwin "was one of about ten major scholars of his day whose writing, public or private, invoked linguistic analogies." (Alter 2002: 4)

Contacts between naturalists and linguists included the close personal and professional relationship between August Schleicher, to whom the first pedigree of Indo-European languages (1853) is attributed, and Ernst Haeckel, the famous populariser of Darwin in Germany. At his inaugural lecture at the University of Bonn in

---

4   Translation from Stevens 1983: 206.

1846, Schleicher spoke about the value of comparative linguistics and argued for language classification based on exact measurements (Schleicher 1850: 25–6). After an extended period of contact with Haeckel, Schleicher published *Die Darwinsche Theorie und die Sprachwissenschaft* (1863), in which he compared the biological concepts of species, sub-species, and variety to the linguistic concepts of language, dialect, and idiom and postulated a "struggle for life" and the presence of "living fossils" among languages. An English edition of the book was published in 1869 under the title *Darwinism Tested by the Science of Language*, indicating an interest in its theme in the English-speaking world. Robert Richards has even called the connections among Darwin, Schleicher, and Haeckel "the missing link in nineteenth-century evolutionary theory" (Richards 2002). Being a cousin of Haeckel the linguist Wilhelm Bleek was also part of this social circle (Koerner 1983:xi). Haeckel contributed a preface to Bleek's *Über den Ursprung der Sprache* (1868) and Bleek for example cited Haeckel's *Generelle Morphologie der Organismen* (1866). He pointed out that Darwin's theory must be corroborated not only through zoology, anatomy, and physiology, but also through geology, archaeology, ethnology, geography, anthropology, and linguistics (Bleek 1867: VII). This shows the universal appeal that evolutionary theory had for some scholars in the 19th century.

As stated above, Darwin on the other hand referred to linguistic evolution and classification, as evolution was firmly established in linguistics by the mid-19th-century. In the first chapter of the *Origin of Species*, Darwin established the similarity between the two theories by writing:

> "[W]e know nothing about the origin or history of any of our domestic breeds. But, in fact, a breed, like a dialect of a language, can hardly be said to have had a definite origin" (Darwin 1859: 40).

In chapter 9 ("On the imperfection of the geological record"), he compared the incomplete geological record to an incomplete body of text:

> "Of this volume, only here and there a short chapter has been preserved, and of each page, only here and there a few lines. Each word of the slowly-changing language, more or less different in the successive chapters, may represent the forms of life, which are entombed in our consecutive formations, and which falsely appear to have been abruptly introduced. On this view the difficulties above discussed are greatly diminished or even disappear" (Darwin 1859: 310–1).

When Darwin established the genealogical classification as the "natural system" in *Origin of Species*, he noted:

> "It may be worth while to illustrate this view of classification, by taking the case of languages. If we possessed a perfect pedigree of mankind, a genealogical arrangement of the races of man would afford the best classification of the various languages now spoken throughout the world; and if all extinct languages and all intermediate and slowly changing dialects, had to be included, such an arrangement would, I think, be the only possible one. Yet it might be that some very ancient language had altered little, and had given rise to few new languages, whilst others (owing to the spreading and subsequent isolation and states of civilisation of the several races, descended from a common race) had altered much, and had given rise to many new languages and dialects. The various degrees of differences in the languages from the same stock, would have to be expressed by groups subordinate to groups; but the proper or even only possible arrangement would still be genealogical; and this would be strictly natural, as it would connect

together all languages, extinct and modern, by the closest affinities and would give the filiation and origin of each language" (Darwin 1859: 422–3).

Some naturalists, such as Haeckel, took up this idea when they suggested a parallel between Indo-Germanic languages and peoples, and generally between languages and human populations (Alter 2002; Römer 1985). Darwin's view that a pedigree of "races" would also provide a classification of languages is rooted in primordialism, i.e. the "assumption of direct linkage between the evolutionary development of languages and 'races' [...] that can be traced back to the turn of the eighteenth to the nineteenth century", and which was the dominant paradigm in linguistics, anthropology, and evolutionary biology during the second half of the 19th century (Kressing et al. 2013: 6).

In *Descent of Man,* Darwin brought his argument full circle to show that his theory of biological evolution was applicable to the cultural domain. He wrote:

"As Max Müller has well remarked: — 'A struggle for life is constantly going on amongst the words and grammatical forms in each language. The better, the shorter, the easier forms are constantly gaining the upper hand, and they owe their success to their own inherent virtue.' To these more important causes of the survival of certain words, mere novelty may, I think, be added; for there is in the mind of man a strong love for slight changes in all things. The survival or preservation of certain favoured words in the struggle for existence is natural selection" (Darwin 1871: 60).

The German-born Friedrich Max Müller had studied the languages of British India for many years and was a professor of linguistics at Oxford University and the most accomplished Indo-Germanist in Britain in the 1870s.

3. FROM SOCIAL CONTACTS TO SOCIAL NETWORK ANALYSIS

References like Darwin's to Müller are not singular, but can be found throughout the formative phase of evolutionary theory in the 19th century. Interdisciplinary contact among scholars played an important role in the development of evolutionary ideas. This "lateral transfer" can be quantified and visualised as a network. In contrast to Alter, who noted that he was "concerned ultimately with large intellectual structures" and took "personal interactions between [...] individuals" (Alter 2002: 109–10), such as Schleicher and Haeckel, only as his starting point, we employ an inductive approach by focusing on such relationships and inferring a larger structure based on that of the emerging network. This network visualises the quantity of interdisciplinary reticulations and is examined here using methods of social network analysis (Wasserman and Faust 1994; Newman 2010).

"Historians, philosophers, and sociologists of science, such as Thomas Kuhn, Robert Merton, and Bruno Latour, have shown that scientific activity and the dissemination of knowledge can be described as social action" (Fangerau 2009: 241). Kuhn used the term "scientific community" to describe a community of scholars who share research interests and methodologies.[5] To reconstruct and map relation-

---

5   See also Fangerau, this volume.

ships among members of the scientific community of evolutionary theorists, we examined available sources documenting actors' references to other actors. We chose 28 actors from the secondary literature on the history of biology and linguistics as the study sample. All but two of them lived in the 19th century and all but one contributed significantly to the scientific discourse of that time. Central individuals in the network were identified by the numbers of network links, which resulted in their centrality on the map of the network. Such visualisation facilitates a synoptic understanding of the relationships among network members, aiding in the process of identifying channels of knowledge transfer. The "visualisation of information aims to reveal insights into complex and abstract information by drawing upon a wide range of perceptual and cognitive abilities of humans" (Chen 2003: 16). By amplifying and reducing elements of the information, it seeks to "provide a means of recognizing patterns and relationships at various levels, which in turn can greatly help us to prioritize where we should search further" (Chen 2003: 16). Thus, the visualisation of information creates an image of the part of the world that is examined, which can cross language barriers, if viewers share the background knowledge needed to interpret it.

Vision is a unique source for cognition. Implicit in every visual perception is knowledge of what underlies the image; sensations and memories join to give the full picture. A picture usually depicts visual items to display "facts". In contrast, an image is composed of several signs (e. g. numbers, symbols, colours, forms) that serve as cultural markers, the connotations of which can be understood like a grammar indicating a certain meaning. To understand images, these codes must be deciphered; they are not self-explanatory or illustrative. This hermeneutic process of "translation" yields an interpretation compatible with the perceiver's world view and cognition. Visual objects in images are designated, rather than denoted, expressing rather than explaining a thesis or model and referring to other sign-relations. In summary, the visualisation of information in images fulfils the classical criterion that a picture is worth a thousand words.[6] An example of these considerations is presented in Figure 2, a word cloud that represents key topics and persons in the social network of evolutionists. The image was artificially prepared using the names and topics considered in this paper. The two central themes of "evolution" and "classification" are large and centrally positioned. The three main disciplines ("biology", "linguistics", and "anthropology") and actors within them are coloured in different shades of grey.[7] The word cloud is arranged like a topical map. It produces the impression, that the whole semantic field in question have might captured at one glance.

6   Chen (2003) discussed these general considerations in greater detail.
7   The social network map below uses the same colour coding.

Figure 2: Themes and actors in the social network of 19th-century evolutionists[8]

Maps effectively visualise information, usually by representing an area of cultural or physical environment. They highlight spatial relationships among elements, such as objects, regions, and concepts within a network. Terrestrial, celestial, and biological maps are classical types. Thematic maps feature overlays that quantitatively or qualitatively depict distribution patterns of specific phenomena. Network phenomena can also be mapped (Chen 2003).[9] Nodes on social network maps represent members and edges between them represent relationships. The whole system of nodes and edges is called a graph. Members of a social network can be closely related via other nodes without being linked directly. A topological representation of the resulting graph that takes into account the strengths of relationships is a map. If scientists are seen as carriers of knowledge, a map of a social network of scientists can depict the topography of knowledge (Chen 2004; Chen 2006).

4. SOURCES, METHODOLOGICAL CHALLENGES, AND CLASSIFICATION OF RELATIONSHIPS

With recognition of the impossibility of reconstructing a complete network of evolutionists based on available historical information, we reconstructed an exemplary network to demonstrate the principle of the approach and foster further research. Thus, our results represent an exploratory attempt and should not be considered final or definitive. The social network was created around central actors taken from the secondary literature on the history of evolution and classification in biology, linguistics, and anthropology. These central characters included Charles Darwin and Ernst Haeckel in biology, August Schleicher and Wilhelm Bleek in linguistics, and Franz Boas and Lewis Henry Morgan in anthropology. Anthropologists were taken into account when it became clear that the discipline had close connections to

---

8   The word cloud was produced by Matthis Krischel with the internet tool available at www.wordle.net (developed by Jonathan Fineberg).
9   Fangerau 2009 explores similar topics in German.

both the study of human languages and human populations and anthropologists offered important contributions to the debate on evolution in the 19th century.

Around these core figures, we established a group of key scholars concerned with evolutionism in the 19th century, as well as some earlier researchers whose work served as points of reference for evolutionary research. Early in the research process, we selected a target group size of about 25 actors to enable the identification of as many relationships between as many actors as possible. The selection of key persons was necessarily biased by the secondary literature; although most members of the network of 19th century scholars in this topic fashionable in Europe at the time were white males, we made a particular effort to identify actors who were not. Nonetheless, all selected actors were from Europe or North America, and Clémence Royer is the only woman among them.[10]

Information about all actors was acquired manually from historical sources without the use of automated, computer-based tools.[11] Special attention was paid to relationships between key persons. For this reason, their number had to be limited, requiring the exclusion of some important contributors to evolutionary discourse, such as the anthropologist Edward Burnett Tylor. Figure 3 shows the life spans of scholars selected as key persons in the network.

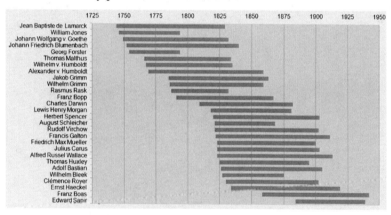

Figure 3: Key persons and their life spans

After key scholars were selected, secondary and primary sources that described their relationships with one another were searched.[12] We started with general, encyclopaedic sources, such as the *Dictionary of Scientific Biography* (Gillispie 1970), and then moved through literature describing the relationship between evolutionary

---

10   The linguist Lucy Lloyd (1834–1914) was included in an early phase of the research because of her working and personal relationship with Bleek, but she was excluded due to the lack of sufficient reticulations with other actors. Further study of her biography and work is necessary to clarify whether this result is due to her poor representation in the secondary literature or due to a lack of relations.

11   The authors would like to acknowledge the support of Frank Kressing and Anja Weigel in data acquisition.

12   A list of the sources consulted is provided in the appendix.

biology and comparative linguistics to works on the histories of individual disciplines and 19th-century science in general and biographies of individual scholars. Marked heterogeneity in the quality and quantity of sources on different actors posed a methodological challenge. The lives and work of famous scientists, such as Darwin, are far better documented than are those of less-known figures. Although large data pools on iconic figures in science, such as the Darwin Correspondence Project, contain letters to less-known actors enabling us to reconstruct relationships to them, we expect that our network is somewhat biased toward actors who are well-represented in the historiography.

We classified relationships into five types: **correspondence relationships**, e.g. letter from Darwin to Haeckel; **citation relationships**, e.g. Darwin cites Müller; **common membership relationships**, e.g. Schleicher and Haeckel were both faculty members at the University of Jena; **personal relationships**, e.g. Bleek was Haeckel's cousin; and **intellectual reference relationships**, e.g. Huxley mentions Virchow in a letter to Darwin. The types of relationship were coded with links to the source in which they were mentioned in a relational Microsoft Access database (figure 4).

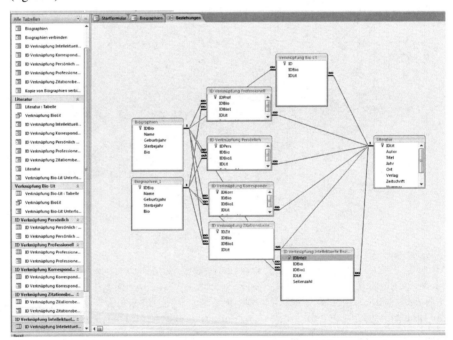

Figure 4: Database structure

When multiple types of relationship between two actors were found, all of them were taken into account and the overall relationship was weighted on a scale ranging from 1 to 5 depending on the number of relationship types.

All correspondence and citation relationships, most common memberships, and some intellectual references were taken from primary sources. Most personal relationships and some intellectual reference relationships were taken from secondary sources. Although the use of primary sources to construct a social network is preferable in most cases, reliance on secondary sources is defensible because they form a body of scholarship produced by generations of historians of science. At the same time, a network based mostly on secondary sources is likely to depict historiographically established connections between historical actors or traditional "narratives".

While gathering data, we faced the challenges of "recall" ("Have I found everything?") and "precision" ("Are my results relevant?") (Rauter 2009). Because data was acquired manually from heterogeneous sources, "recall" was a much greater challenge, and the likelihood that not all relevant relationships between actors were found must be recognised. Fortunately, this process has high "precision", with very few "false positive" results expected. The most likely source for such results is the presentation of so-called "historical myths" in the secondary literature.

A total of 223 relationships among the 28 actors in the sample were identified: 33 correspondence, 61 citation, 54 common membership, 34 personal, and 41 intellectual reference relationships. These relationships include "formal" and "informal" modes of exchange (Fangerau 2009: 218–24; cf. Fangerau this volume). Formal modes of exchange are public references according to scientific and, often, disciplinary norms, including citations and intellectual references in print or public speeches. Informal modes of exchange, which formed scholars' intellectual bases and informed their thought processes, were also examined to facilitate the understanding of scientists as real-life people and science as social practice.

5. RESULTS: A SOCIAL NETWORK OF 19TH-CENTURY EVOLUTIONISTS

The NodeXL software[13] was used to quantify and visualise a map of the social network (Smith et al. 2009). Multidimensional scaling was used to depict the network as clearly and in as visually appealing a manner as possible. Node sizes correspond to the total numbers of other nodes to which they connect (i.e. the number of relationships with other actors), termed "degree centrality" in social network analysis. Node positions on the map were determined by their "betweenness centrality" which measures the shortest paths from a node to other nodes in the network. Thus, actors positioned centrally on the map tended to be well connected or to have relationships with well-connected actors (figure 5).

Although interdisciplinary research was common during the time period under consideration, nodes are colour coded according to actors' currently recognised primary disciplinary affiliations. Different shades of grey are used for biologists, linguists, and anthropologists. Two actors, Thomas Malthus and Johann-Wolfgang von Goethe, served as important intellectual references for numerous other members of the network but could not be assigned to one of these categories and are thus

---

13   http://nodexl.codeplex.com [last access: 2013–05–13].

depicted in yet another grey. Relationships between actors can be deduced by the positions of nodes on the map.

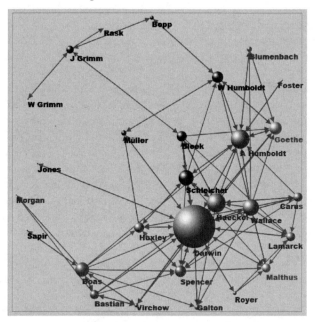

Figure 5: Social network map of 19th-century evolutionists

5.1. Interpretation of the mapped network

Darwin is obviously (and not surprisingly, given our selection of sources) the central actor in the network. He is represented by the largest node, which is located at the centre of the map. This result can be explained by several factors. Darwin is well represented in histories of biology and in secondary literature concerned with the reticulations between biology and linguistics and between biological and cultural anthropology in the 19th century. Secondary sources also indicate that colleagues from Britain, Europe, North America and Asia considered Darwin to be "well-connected" during his lifetime. He maintained these connections largely through a correspondence network; Darwin exchanged about 15,000 letters with nearly 2,000 correspondents during his lifetime, many of which can be found at the Darwin Correspondence Project's website.[14] Haeckel's correspondence network was even larger; nearly 40,000 letters to and from him are archived in Jena and await further study (Hoßfeld and Breidbach 2005). Correspondence networks like those of Darwin and Haeckel illustrate the scale of informal exchange of specimens, research data, ideas, and opinions in the 19th century. These networks can be argued to have been arenas of peer review prior to publication, most often of mono-

14   cf. http://www.darwinproject.ac.uk/ [last access: 2013–05–01].

graphs, a role similar to that of preprints in some scientific disciplines today. Letters also often contain references to a third party with whom both correspondents are in contact, which forms part of their intellectual bases or is part of public discourse (Fangerau 2009: 223).

Visual analysis reveals that the network consists of three clusters, defined in social network analysis as groups of nodes connected more closely to one another than to the rest of the network. In the network of 19th-century evolutionists, these clusters signify groups of scholars who were in closer contact with one another than with the rest of the community. The first cluster is arranged around Franz Boas and contains his pupils, the linguist Edward Sapir and anthropologist Adolf Bastian, and the anthropologists Lewis Henry Morgan and Rudolf Virchow. The second cluster is centred around Jakob Grimm and includes his brother Wilhelm, Rasmus Rask, and Franz Bopp, all of whom were linguists. The third, central cluster around Darwin is markedly more interdisciplinary, comprised of biologists, linguists, anthropologists, and two scholars classified as "other". The map indicates that this central group of scholars contributed most to the formulation of evolutionary theory and classification based on common descent in the 19th century.

Following Ludwik Fleck, this central cluster can be termed the "esoteric circle" of evolutionism. Fleck noted that members of such a circle within a thought collective are connected by a "solidarity of thought" (Fleck 1980: 140); in the present case, evolutionism and classification based on common descent. Analysis of the social network map indicates that this characterisation is most applicable to the most central actors: Darwin, Spencer, Huxley, Wallace, Haeckel, and Schleicher. The map displays Fleck's notion of an esoteric circle within a thought collective, but does not clearly demarcate an "exoteric circle". Instead, inclusion in the thought collective of evolutionism appears to diminish gradually with distance from the network's centre. The central cluster contains three individuals who were significantly older than Darwin (Jean-Baptiste de Lamarck, Robert Malthus, and Alexander von Humboldt) and served primarily as intellectual references for cluster members, who were around the same age and were scientifically active at the same time.

Social network analysis distinguishes three special kinds of nodes: landmark, hub, and pivot nodes (Chen 2004). In the mapped social network of 19th-century evolutionists, the radius of landmark nodes is large, signifying the great "weight" of relationships. In this network, Darwin is signified by the only obvious landmark node, indicating that he had the largest number of strongest relationships with other actors. Hub nodes signify a large number of relationships with other actors, which are not necessarily strong. They represent well-connected actors; in this network, Darwin, Huxley, Spencer, Wallace, Haeckel, Schleicher, and Humboldt. Actors may be signified by hub nodes for different reasons. Alexander von Humboldt, for example, was a main authority in natural history and was famous for his voyages of exploration. For these reasons, he was in contact with many network members or contributed to their intellectual bases. Thomas Huxley, on the other hand, was not only known as "Darwin's bulldog", i.e. his populariser even in controversial situations, but he also read German, helping to connect Darwin to German-language literature. The network contains few pivot nodes. Four actors are connected to the network through only one other actor (Sapir through Boas, Rask and Wilhelm Grimm through Jakob Grimm, and Foerster through Goethe), they appear as "brokers" or connectors in this network, although extensive research would likely reveal additional connections, leading to the conversion of these pivot nodes to another node type. This result is mainly due to the characteristics of key actors in this sample, all of whom were renowned and highly regarded scholars.

6. CONCLUSION

The construction of a social network from historical sources and its computer-aided visualisation provide many advantages. Visual representation makes obvious the many reticulations among naturalists, linguists, anthropologists, and other scholars who contributed to the formulation of evolutionary theory and classification based on common descent in the 19th century. The network map also clearly shows that interdisciplinary contact was not an exception, but rather the rule among those scholars.

However, the network map is to a degree a representation of current scholarly writing about evolutionary theory, given the sources on which it was based. Despite the use of primary sources, it mainly represents connections that other authors have considered sufficiently meaningful to mention. Nevertheless, these authors did not necessarily seek to describe an interdisciplinary network; thus, our approach depicts this data in a new light. Another advantage of our approach is the inclusion not only of obviously "central" actors in the historiography of evolution and classification, but also contacts who are less well known today. This advantage can be reduced by the collection of data from secondary sources, which represent the historiography, and increased by the incorporation of more primary sources, including correspondence corpora. The central positions of some actors in the network, including Darwin and Haeckel, can be explained by the practice of letter writing, an important means of scientific communication in the 19th century.[15] Both of these scholars cultivated extensive correspondence networks and benefited from the ability to rely on numerous contacts. At the same time, personal contacts, e.g. those among Haeckel, Schleicher, and Bleek, played a surprisingly important role in determining the topology of the network map. Historians of science should further examine the roles of these personal relationships, which fostered interdisciplinary contact.

Finally, the mapped network helps us to pose further questions about interdisciplinary development of scientific knowledge. The use of our approach to analyse an expanded study sample with additional archival materials might reveal the "co-evolution" and "lateral transfer" of ideas from a diachronic perspective, thereby offering further insight into processes of innovation in science that might be valuable for current researchers eager to transgress disciplinary boundaries.

---

15   For a recent example, cf. Fangerau 2010.

REFERENCES

Alter, S. (2002) Darwinism and the Linguistic Image: Language, Race, and Natural Theology in the Nineteenth Century (Baltimore: Johns Hopkins University Press)

Archibald, J.D. (2009) 'Edward Hitchcock's Pre-Darwinian (1840) "Tree of Life"', Journal of the History of Biology 42(3) (1. August 2009): 561–92. doi:10.1007/s10739–008–9163-y.

Bleek, W. (1867) Über den Ursprung der Sprache (Kapstadt: Manuskript).

Bock, K.E. (1955) 'Darwin and Social Theory', Philosophy of Science 22(2): 123–34.

Bowler, P.J. (2003) Evolution: The History of an Idea (Berkeley: University of California Press).

Casanova, P. (2004) The World Republic of Letters (Cambridge: Harvard University Press).

Chen, C. (2003) Mapping Scientific Frontiers: The Quest for Knowledge Visualisation (London: Springer).

Chen, C. (2004) 'Searching for intellectual turning points: Progressive knowledge domain visualization', Proceedings of the National Academy of Sciences 101(suppl. 1) (23. January 2004): 5303–10. doi:10.1073/pnas.0307513100.

Chen, C. (2006) 'CiteSpace II: Detecting and visualizing emerging trends and transient patterns in scientific literature', J. Am. Soc. Inf. Sci. Technol. 57(3) (February 2006): 359–77. doi:10.1002/asi.v57:3.

Dalton, S. (2003) Engendering the Republic of Letters: Reconnecting Public and Private Spheres. (Montreal: McGill-Queen's University Press).

Darwin, C. (1859) On the Origin of the Species by Means of Natural Selection: Or, The Preservation of Favoured Races in the Struggle for Life (London: John Murray).

Darwin, C. (1871) The Descent of Man (New York: D. Appleton and Company).

Fangerau, H. (2009) 'Der Austausch von Wissen und die rekonstruktive Visualisierung formeller und informeller Denkkollektive', in H. Fangerau & T. Halling (eds), Netzwerke: allgemeine Theorie oder Universalmetapher in den Wissenschaften? (Bielefeld: transcript Verlag): 215–46.

Fangerau, H. (2010) Spinning the scientific web: Jacques Loeb (1859–1924) und sein Programm einer internationalen biomedizinischen Grundlagenforschung (Berlin: Akademie Verlag).

Farber, P. (2000) Finding Order in Nature: The Naturalist Tradition from Linnaeus to E.O. Wilson (Baltimore: Johns Hopkins University Press).

Fleck, L. (1980) Entstehung und Entwicklung einer wissenschaftlichen Tatsache: Einführung in die Lehre vom Denkstil und Denkkollektiv (Berlin: Suhrkamp).

Gillispie, C.C. (ed) (1970) Dictionary of Scientific Biography (New York: Scribner).

Hoßfeld, U. & O. Breidbach (2005) Haeckel-Korrespondenz: Übersicht über den Briefbestand des Ernst-Haeckel-Archivs (Berlin: Verlag für Wissenschaft und Bildung).

Koerner, K. (1981) 'Schleichers Einfluß auf Haeckel: Schlaglichter auf die wechselseitige Abhangigkeit zwischen linguistichen und biologischen Theorien in 19. Jahrhundert', Zeitschrift für vergleichende Sprachforschung 91: 1–21.

Koerner, K. (1983) Linguistics and Evolutionary Theory. Three Essays by August Schleicher, Ernst Haeckel, and Wilhelm Bleek (Amsterdam, Philadelphia: John Benjamins)

Kressing, F.; Krischel, M. & H. Fangerau (2013) 'The "Global Phylogeny" and Its Historical Legacy: A Critical Review of a Unified Theory of Human Biological and Linguistic Co-Evolution', Medicine Studies: 1–13. doi:10.1007/s12376–013–0081–8.

Krischel, M.; Kressing, F. & H. Fangerau (2011) 'Die Entwicklung der Deszendenztheorie in Biologie, Linguistik und Anthropologie als Austauschprozess zwischen Geistes- und Naturwissenschaften', in H.-K. Keul & M. Krischel (eds), Deszendenztheorie und Darwinismus in den Wissenschaften vom Menschen, (Stuttgart: Steiner-Verlag): 107–22.

Kuhn, T.S. (1970) The structure of scientific revolutions. 2nd, enl. International encyclopedia of unified science. Foundations of the unity of science (Chicago: University of Chicago Press).

Larson, E. (2006) Evolution: The Remarkable History of a Scientific Theory (New York: Random House Publishing).

Lockwood, W. B. (1969) Indo-European Philology: Historical and Comparative (London: Hutchinson).

Mauelshagen, F. (2003) 'Networks of Trust: Scholarly Correspondence and scientific exchange in early modern Europe', *The Medieval History Journal* 6: 1–32.

Mauro, T. de; Formigari, L. & S. Aroux (eds) (1990) 'Representation and Place of Linguistic Change before Comparative Grammar', in T. De Mauro & L. Formigari, Leibniz, Humboldt and the Origins of Comparativism [papers in Revised Versions of the Actual Presentations at the Conference on Leibniz, Humboldt, and the Origins of Comparativism Held in Villa Mirafiori, Roma, Sept. 1986] (Amsterdam, John Benjamins Publishing): 231–8.

Newman, M. (2010) Networks: An Introduction (Oxford: Oxford University Press).

Pearl, J.L. (1984) 'The Role of Personal Correspondence in the Exchange of Scientific Information in Early Modern France', Renaissance and Reformation 20: 106–13.

Rauter, J. (2009) 'Textvernetzungen und Zitationsnetzwerke', in H. Fangerau & T. Halling (eds), Netzwerke: allgemeine Theorie oder Universalmetapher in den Wissenschaften? (Bielefeld: transcript Verlag): 247–66.

Richards, R. (2002) 'The Linguistic Creation of Man: Charles Darwin, August Schleicher, Ernst Haeckel, and the Missing Link in Nineteenth-Century Evolutionary Theory', in M. Dörries (ed), In Experimenting in Tongues: Studies in Science and Language (Stanford: Stanford University Press): 21–48.

Römer, R. (1985) Sprachwissenschaft und Rassenideologie in Deutschland (München: W. Fink).

Rusnock, A. (1990) 'Correspondence networks and the Royal Society, 1700–1750', The British Journal for the History of Science 32: 155–69.

Schleicher, A. (1850) 'Über den Werth der Sprachvergleichung', Bonner Antrittsvorlesung vom 27. Juni 1846, Lassens Zeitschrift für die Kunde des Morgenlandes 7:25–47.

Secord, J.A. (2000) Victorian Sensation: The Extraordinary Publication, Reception, and Secret Authorship of Vestiges of the Natural History of Creation (Chicago: University of Chicago Press).

Smith, M.A.; Shneiderman, B.; Milic-Frayling, N.; Mendes Rodrigues, E.; Barash, E.; Dunne, C.; Capone, T.; Perer, A. & A. Gleave (2009) 'Analyzing (social media) networks with NodeXL', Proceedings of the fourth international conference on Communities and technologies, 255–264. University Park, PA, USA: ACM, 2009.

Steinke, H. & M. Stuber (2004) 'Medical Correspondence in Early Modern Europe. An Introduction', Gesnerus 61: 139–60.

Stevens, P.F. (1983) 'Augustin Augier's "Arbre Botanique" (1801), a Remarkable Early Botanical Representation of the Natural System', Taxon 32(2) (Mai 1983): 203. doi:10.2307/1221972.

Taub, L. (1993) 'Evolutionary ideas and 'empirical' methods: the analogy between language and species in works by Lyell and Schleicher', The British Journal for the History of Science 26(2): 171–93. doi:10.1017/S0007087400030740.

Trigger, B.G. (1998) Sociocultural Evolution: Calculation and Contingency (Weinheim: Wiley).

Wasserman, S. & K. Faust (1994) Social Network Analysis: Methods and Applications (Cambridge: Cambridge University Press).

APPENDIX: LIST OF SOURCES CONSULTED
TO CONSTRUCT THE SOCIAL NETWORK

The sources listed in this appendix were used in addition to those listed in the chapter bibliography. All internet sources were last checked 1 May 2013.

Letters	
	A v. Humboldt to Darwin, Letter 534 (1839), http://www.darwinproject.ac.uk/entry-534
	Darwin to A v. Humboldt, Letter 545 (1839), http://www.darwinproject.ac.uk/entry-545
	Darwin to Carus, Letter 5273 (1866), http://www.darwinproject.ac.uk/entry-5273
	Darwin to Carus, Letter 5282 (1866), http://www.darwinproject.ac.uk/entry-5282
	Darwin to Carus, Letter 6513 (1866), http://www.darwinproject.ac.uk/entry-6513
	Darwin to Carus, Letter 5375 (1867), http://www.darwinproject.ac.uk/entry-5375
	Darwin to Carus, Letter 5403 (1867), http://www.darwinproject.ac.uk/entry-5403
	Darwin to Galton, Letter 1525 (1853), http://www.darwinproject.ac.uk/entry-1525
	Darwin to Galton, Letter 1881 (1854), http://www.darwinproject.ac.uk/entry-1881
	Darwin to Galton, Letter 1627 (1855), http://www.darwinproject.ac.uk/entry-1627
	Darwin to Galton, Letter 3059 (1856), http://www.darwinproject.ac.uk/entry-3059
	Darwin to Galton, Letter 2121 (1857), http://www.darwinproject.ac.uk/entry-2121
	Darwin to Galton, Letter 2581 (1859), http://www.darwinproject.ac.uk/entry-2581
	Darwin to Galton, Letter 2799 (1860), http://www.darwinproject.ac.uk/entry-2799
	Darwin to Galton, Letter 5800 (1868), http://www.darwinproject.ac.uk/entry-5800
	Carus to Darwin, Letter 5269 (1866), http://www.darwinproject.ac.uk/entry-5269
	Carus to Darwin, Letter 5279 (1866), http://www.darwinproject.ac.uk/entry-5279
	Carus to Darwin, Letter 5370 (1867), http://www.darwinproject.ac.uk/entry-5370
	Carus to Darwin, Letter 5397 (1867), http://www.darwinproject.ac.uk/entry-5397
	Carus to Darwin, Letter 5489 (1867), http://www.darwinproject.ac.uk/entry-5489
	Galton to Darwin, Letter 2573 (1959), http://www.darwinproject.ac.uk/entry-2573
	Galton to Darwin, Letter 7026 (1969), http://www.darwinproject.ac.uk/entry-7026
	Royer to Darwin, Letter 5339 (1865), http://www.darwinproject.ac.uk/entry-5339
	Darwin to Wallace, Letter 2087 (1857), http://www.darwinproject.ac.uk/entry-2086
	Wallace to Darwin, Letter 2145 (1858), http://www.darwinproject.ac.uk/entry-2145
	Müller to Darwin, Letter 8957 (1873), http://www.darwinproject.ac.uk/entry-8957
	Darwin to Müller, Letter 9802 (1875), http://www.darwinproject.ac.uk/entry-9802
	Bastian to Virchow (1889), http://www.bgaeu.de/BGAEU-AUT.htm
	Darwin an Virchow (1878–1888), http://www.bgaeu.de/BGAEU-AUT.htm
	Virchow to Darwin (1878–1888), http://www.bgaeu.de/BGAEU-AUT.htm

Books, chapters and journal articles used as primary sources in chronological order	Malthus, Thomas. An Essay on the Principle of Population as it Affects the Future Improvement of Society, with Remarks on the Speculations of Mr. Godwin, M. Condorcet, and other Writers. 1798. London, J. Johnson.
	Lyell, Charles. Principles of geology. An attempt to explain the former changes of the earth's surface, by reference to causes now in operation. 1830. London, John Murray.
	Schleicher, August. Die ersten Spaltungen des indogermanischen Urvolkes. 1853. Allgemeine Monatsschrift für Wissenschaft und Literatur.
	Wallace, Alfred Russel. On the Rio Negro. 1853. Journal of the Royal Geographical Society 23, 212–217.
	Spencer, Herbert. Progress: It´s Law and Cause. 1857. London, Williams and Norgate.
	Darwin, Charles Robert. On the Origin of Species by Means of Natural Selection, or the Preservation of Favoured Races in the Struggle for Life. 1859. London, John Murray.
	Bronn, Heinrich. Über die Entstehung der Arten im Thier- und Pflanzen-Reich durch natürliche Züchtung, oder Erhaltung der vervollkommneten Rassen im Kampfe um's Daseyn. 1860. Stuttgart, E. Schweizerbart'sche Verlagshandlung und Druckerei.
	Grimm, Jacob. Über den Ursprung der Sprache. 5. Aufl. 1862. Berlin, Dümmler.
	Wallace, Alfred Russel. On the Varieties of Men in the Malay Archipelago (a paper read in Newcastle-upon-Tyne at the 1 Sept. 1863 meeting of Section E, Geography and Ethnology, of the BAAS). 1863.
	Wallace, Alfred Russel. On the Geographical Distribution of Animal Life (a paper read in Newcastle-upon-Tyne at the 31 Aug. 1863 meeting of Section D, Zoology and Botany, of the BAAS). 1863.
	Spencer, Herbert. The Principles of Biology. 2 volumes. 1864–1867. New York, Apleton.
	Schleicher, August. Ueber die Bedeutung der Sprache für die Naturgeschichte des Menschen. 1865. Weimar.
	Haeckel, Ernst. Generalle Morphologie der Organismen. Allgemeine Grundzüge der organischen Formen-Wissenschaft, mechanisch begründet durch die von Charles Darwin reformierte Descendenztheorie. 1866. Berlin, Reimer.
	Haeckel, Ernst. Vorwort zu Bleek, Wilhelm: Der Ursprung der Sprache. 1868. Weimar, Böhlau
	Carus, Julius Victor. Das Variieren der Thiere und Pflanzen im Zustande der Domestication. 1868. Stuttgart, E. Schweizerbart´sche Verlagshandlung.
	Bleek, Wilhelm. Über den Ursprung der Sprache. 1868. Weimar, Böhlau.
	Haeckel, Ernst.Über die Entstehung und den Stammbaum des Menschengeschlechts. 1868. Berlin
	Wallace, Alfred Russel. Discussion [of paper on the races and antiquity of mankind by T.H. Huxley read in Norwich at the 24 Aug. 1868 "Fourth Meeting" of the Third International Congress of Prehistoric Archæology]. 1869. London, Longmans, Green, & Co.
	Wallace, Alfred Russel. Review of Hereditary Genius: An Inquiry Into 1st Laws and Consequences by Francis Galton, 1869. Nature. 1870.

Darwin, Charles (transl. by Carus, Julius Victor). Die Abstammung des Menschen und die geschlechtliche Zuchtwahl. 1871. Stuttgart, E. Schweizerbart´sche Verlagshandlung (E. Koch).

Darwin, Charles (transl. by Carus, Julius Victor). Der Ausdruck der Gemüthsbewegungen bei dem Menschen und den Thieren. 1872. Stuttgart, E. Schweizerbart´sche Verlagshandlung (E. Koch).

Carus, Julius Victor. Alexander von Humboldt. Eine wissenschaftliche Biographie. 1872. Leipzig.

Wallace, Alfred Russel. The President's Address. Proceedings of the Entomological Society of London for the Year 1871. 1872.

Müller, Max. Max Muller on Darwin's Philosophy of Language. Nature 7, 145. 1872.

Schleicher, August. Die Darwinsche Theorie und die Sprachwissenschaft. Offenes Sendschreiben an Herrn Dr. Ernst Haeckel, o.Professor der Zoologie und Director des zoologischen Museums an der Universität Jena. 1873. Weimar, Hermann Böhlau.

Wallace, Alfred Russel. Review of Advanced Text-book of Physical Geography (2nd ed.) by David Page. Nature 201, 358–361. 1873.

Spencer, Herbert. Principles of Sociology. 1874. New York, Appleton.

Carus, Julius Victor. Charles Darwins gesammelte Werke. Autorisierte deutsche Ausgabe. 1874. Stuttgart, E. Schweizerbart'sche Verlagshandlung (E. Koch).

Darwin, Charles. Reise eines Naturforschers um die Welt. 1875. Stuttgart, E. Schweizerbart'sche Verlagshandlung (E. Koch).

Darwin, Charles (transl. by Carus, Julius Victor).Insectenfressende Pflanzen. 1876. Stuttgart, E. Schweizerbart'sche Verlagshandlung (E. Koch).

Wallace, Alfred Russel. The Geographical Distribution of Animals; With A Study of the Relations of Living and Extinct Faunas as Elucidating the Past Changes of the Earth's Surface. 1876. London, Macmillan & Co.

Carus, Julius Victor. Die verschiedenen Blüthenformen an Pflanzen der nämlichen Art. 1877. Stuttgart, E. Schweizerbart'sche Verlagshandlung und Druckerei (E. Koch).

Carus, Julius Victor. Die Wirkungen der Kreuz- und Selbst-Befruchtung im Pflanzenreich. 1877. Stuttgart, E. Schweizerbart'sche Verlagshandlung und Druckerei (E. Koch).

Wallace, Alfred Russel. Review of Evolution, Old and New; Or, The Theories of Buffon, Dr. Erasmus Darwin, and Lamarck, As Compared With That of Mr. Charles Darwin. Nature 20, 141–144. 1879.

Wallace, Alfred Russel. Review of The Evolution of Man: A Popular Exposition of the Principal Points of Human Ontogeny and Phylogeny by Ernst Haeckel. Academy 15, 326–327, 351–352. 1879.

Carus, Julius Victor. Das Bewegungsvermögen der Pflanzen. 1881. Stuttgart, E. Schweizerbart'sche Verlagshandlung und Druckerei (E. Koch).

Carus, Julius Victor. Die Bildung der Ackererde durch die Thätigkeit der Würmer. 1882. Stuttgart, E. Schweizerbart'sche Verlagshandlung und Druckerei (E. Koch).

Spencer, Herbert (tranls. by Victor Carus). Die Principien der Ethik. 1895. Stuttgart, E. Schweizerbart'sche Verlagshandlung (E. Koch)

	Spencer, Herbert (tranls. by Victor Carus). Grundsätze einer synthetischen Auffassung der Dinge. 1900. Stuttgart, E. Schweizerbart'sche Verlagshandlung (E. Koch). Wallace, Alfred Russel. The Passing Century – Evolution. The Sun 68 (114), 4a-g, 5a. 1900. Wallace, Alfred Russel. My Life an autobiography. 1905. London, Chapman & Hall. Wallace, Alfred Russel. Note on the Passages of Malthus's 'Principles of Population' Which Suggested the Idea of Natural Selection to Darwin and Myself. 1909. London, Burlington House, Longmans, Green & Co. Lloyd, Lucy. Specimens of Bushman Folklore. 1911. London, G. Allen. Boas, Franz. Kultur und Rasse, zweite unveränderte Auflage. 1922. Berlin, Leipzig, Walter de Gruyter & co. Leitzmann, Albert. Gesammelte Schriften von Wilhelm von Wilhelm von Humboldt. 1903–1936, reprint 1968. Berlin, Preußische Akademie der Wissenschaften. Sapir, Edward. Sound Patterns in Language. In: Selected Writings of Edward Sapir on Language, Culture, and Personality, ed. by David Mandelbaum. 1949. Berkeley, University of California Press. Sapir, Edward. The Unconsious Patterning of Behaviour in Society. In: Selected Writings of Edward Sapir on Language, Culture, and Personality, ed. by David Mandelbaum. 1949. Berkeley, University of California Press. Boas, Franz. Race, Language, Culture. 1949. New York. Andree, Christian. Rudolf Virchow, Gesammelte Werke. Abt IV (Briefe). 1992–2010. Hildesheim, Georg Olms. Darwin, Charles (tranls. by Victor Carus). Reise um die Welt: 1831–36. 1993. Stuttgart, Wien, Thienemann.
Biographical Compendia in chronological order	Hirsch, August (ed). Biographisches Lexikon der hervorragenden Ärzte aller Zeiten und Völker. 1962. München, Urban und Schwarzenberg. Gillispie, Charles (ed). Dictionary of Scientific Biography. 1970–1980. New York, Charles Scribner's Sons. Koertge, Noretta (ed). New Dictionary of Scientific Biographies. 2007. Charles Scribner's Sons. Darwin, Charles. Gesammelte Werke. 2008. Frankfurt am Main.
Books, chapters and journal articles used as secondary sources in chronological order	Steinthal, H. Die sprachphilosophischen Werke Wilhelm von Humboldts. 1884. Berlin. Francis Darwin (ed). The Life and Letters of Charles Darwin, including an autobiographical chapter. 1887. London, John Murray. Duncan, David. Life and Letters of Herbert Spencer. 1908. New York, D. Appleton. Maher, J. Introduction in Linguistics and Evolutionary Theory. 1983. Amsterdam, J. Benjamins. Korner, Konrad. Linguistics and Evolutionary Theory. Three Essays by August Schleicher, Ernst Haeckel, and Wilhelm Bleek. 1983. Amsterdam, John Benjamins.

Hull, David. Science as a Process: An Evolutionary Account of the Social and Conceptual Development of Science. 1988. Chicago, London, University of Chicago Press.

Römer, Ruth. Sprachwissenschaft und Rassenideologie in Deutschland. 1989. München, Wilhelm Finck.

Andree, Christian. Rudolf Virchow, Gesammelte Werke. Abt. I (Medizin), III (Anthropologie, Ethnologie). 1992–2010. Hildesheim, Georg Olms.

Harvey, Joy. Almost a Man of Genius: Clémence Royer, Feminism and Nineteenth-Century Science. 1997. New Brunswick, New Jersey, Rutgers University Press.

Alter, Stephen. Darwinism and the linguistic Image: Language, Race, and Natural Theology in the Nineteenth Century. 1999. London, Baltimore, Johns Hopkins University Press.

Streck, Bernard. Kulturanthropologie. 2000. Wuppertal, Hammer-Verlag.

Streck, Bernard. Diffusion. 2000. Wuppertal, Hammer-Verlag.

Gingrich, Andre. The German-Speaking Countries. In: One Discipline, Four Ways: British, German, French, and American Anthropology, ed. by C. Hann. 2005. Chicago, London, University of Chicago Press.

Parkin, Robert. The French-Speaking Countries. In: One Discipline, Four Ways: British, German, French, and American Anthropology, ed. by C. Hann. 2005. Chicago, London, University of Chicago Press.

Pöhl, Friedrich and Bernhard Tilg. Franz Boas. Kultur, Sprache, Rasse. Wege einer anti-rassistischen Anthropologie. 2009. Wien, Berlin, Münster, Lit.

Atkinson, Quentin and Russel Gray. Curious Parallels and Curious Connections – Phylogenetic Thinking in Biology and Historical Linguistics. Systematic Biology 54 (4), 517. 2005.

Krischel, Matthis. Perceived Hereditary Effect of World War I: A Study of the Positions of Friedrich von Bernardi and Vernon Kellog. Medicine Studies 2 (2), 139–150. 2010.

TRANSLATING NATURAL SELECTION:
TRUE CONCEPT, BUT FALSE TERM?

Thierry Hoquet

INTRODUCTION

It is now well known that Darwin's (1859) theory of *"the origin of species by means of natural selection"* was met with strong opposition (Bowler 1983). Some theoretical issues with natural selection were at stake. For instance, the original debate between Darwin and Wallace shows quite clearly that the main point of contention between the two "co-discoverers" of natural selection was that selection operated on the level of individual variations, rather than varieties (or "races"), and therefore was believed not to produce sufficiently stable results (Gayon 1998). This paper does not challenge this view, but aims at stressing some linguistic aspects of the same issue. Under attack was not only the power of natural selection to account for the transformation of species; Darwin also had to face critiques of the phrase itself. The outrage was in fact rather universal, and the choice of the term *"natural selection"* was constantly questioned in readers' reviews and comments. Darwin's outright opponent Bishop Wilberforce puts it quite clearly:

> "Nor must we pass over unnoticed the transference of the argument from the domesticated to the untamed animals. Assuming that man as the selector can do much in a limited time, Mr. Darwin argues that Nature, a more powerful, a more continuous power, working over vastly extended ranges of time, can do more. But why should Nature, so uniform and persistent in all her operations, tend in this instance to change? why should she become a selector of varieties?" (Wilberforce 1860: 237).

As the historian Robert Young stated, it seems the main question posed by Darwin's readers was: *"does nature select?"* (Young 1985). Facing these attacks, Darwin constantly receded. Summarizing the fourth chapter in the first edition, he made clear that *"natural selection"* was just a way to encapsulate *"for the sake of brevity"*, a quite complex *"principle of preservation"* (Darwin 1859: 127). As early as the second edition (December 1859), he specified that it is only *"metaphorically"* that natural selection can be said to be *"daily and hourly scrutinising, throughout the world, every variation, even the slightest"* (as the first edition initially read) (Darwin 1959: 169)[1]. A few months later, in the third edition (April 1861), Darwin

---

1   For practical reasons, all references to the various English editions of Darwin's *Origin* are to the *Variorum Edition* edited by Morse Peckham (Darwin 1959). Since this impressive work is currently viewed with increasing suspicion, quotations have been systematically checked on the original texts, available on line, thanks to the website http://darwin-online.org.uk, edited by John Van Whye. Even ways of quoting are subject to fashion.

bluntly admitted that natural selection was, "*no doubt*", a "*misnomer*". In 1868, Darwin made clear, in the introduction to his *Variation under domestication*, that "*the term 'natural selection' is in some respects **a bad one**, as it seems to imply conscious choice; but, Darwin hopes, this will be disregarded after a little familiarity*" (Darwin 1868: 6; emphasis added). In the fifth edition of the *Origin* (August 1869), he finally went as far as to call natural selection a "*false term*" and appends to it the Spencerian "*survival of the fittest*" (Darwin 1959: 164–5), pairing them for several decades as the Tweedledum and Tweedledee of evolutionary biology. Darwin's position was very clear: "*natural selection*" is a bad term for a sound concept. But can one really sever the inappropriate term from the great idea?

We have to understand why, while acknowledging the shortcomings of his original phrase, Darwin nonetheless maintained it as a key concept in the architecture of his theory and did not simply give it up. Focusing on the case of "*natural selection*", this essay develops an approach to Darwin's scientific vocabulary from the perspective of the history and philosophy of science. It analyses the cases of French and German translations not to criticize them but to see how they shed light on the complex original English term. I use translations as a prism that diffracts the white light emanating from the *Origin* and helps us improve our understanding of Darwin's intentions and aims (Hoquet 2009).

A CAUTIONARY TALE ON DOGS AND BANANAS

What can we learn from translations about the exportation of Darwin's *Origin of species*[2]? Natural selection is clearly a cornerstone for understanding the Darwinian ideas of nature and evolution, several aspects of which have been studied so far. Considerable attention has been devoted to the *concept* of natural selection and detailed accounts of this mechanism and its philosophical upshot have been developed (Sober 1984; Huneman 2009). Some scholars have tried to give us a better overview of the *comparative reception* of the Darwinian theory (Ellegård 1958/1990; Glick 1974; and relevant chapters in Kohn 1985). Others have analysed Darwin's path to his *discovery* of natural selection, weighing the importance of different factors or fields, such as biogeography, artificial selection, Darwin's study of the barnacles or the theory of generation (Limoges 1970; Ruse 1975; and Sloan and Hodge's chapters in Kohn 1985). The question of the units or levels of selection has developed as a category of its own in the field of philosophy of biology (Brandon

---

2   Due to the author's own linguistic limitations, this paper will focus on a few West European languages: mostly French and German, with some quick incursions into Dutch, Italian or Spanish. This might be relevant for the study of the fate of Darwinian terms, since those were among the first languages into which the *Origin* was translated and where "*natural selection*" was liable to transliteration. Undoubtedly, the analysis should be further pursued and its conclusions confronted to wider linguistic horizons. See, e.g., for the translations in Arabic, Elshakry (2008), especially for the tension between derivation and Arabicization, or transliteration and neologism, and its political significance; for the translation of Darwin in Japanese, see Montgomery (2000: 232–235); for Russian, see Todes (1989).

and Burian 1984; Sober and Wilson 1998; Keller 1999; Okasha 2006). But the *term* "*natural selection*" has received far less attention, too little attention in fact, with the notable exception of Hodge (1992).

Moreover, until quite recently (Montgomery 2000), scientific translations have not been a focus of much scholarly interest. As was aptly noted by Nicolaas Rupke (2000: 209), "*the assumption has been widespread that the language of science was, and continues to be, an international language, a* lingua franca". At most, translations have long been considered a good but merely bibliometric indicator of the success of a book, or a measure of the diffusion and reception of a theory. More recently however, it seems that the various renditions of scientific texts have been considered a tool for the historian of science to get a grasp on more local traditions. Processes of transfer have been re-read as assimilations, revealing not only the impact of the Original on a local context but also the retroaction of the autochthons' minds on the Original itself. As was already known in the seventeenth century with Nicolas Perrot d'Ablancourt, translations are often *belles infidèles* rather than *faithful* renderings (Zuber 1968). All this, no doubt, opens interesting debates on contingencies in the history of ideas and seamlessly connects with recent emphasis on the situatedness of knowledge (Haraway 1988). But it would eventually amount to no more than ceaseless scholarly embroidery on the Latin phrase "*omnis traductor traditor*", "*every translator is a traitor*", with some additional postmodern localist seasoning. The moral of the story would be that translating is corrupting, that the translator should be faithful and invisible (Venuti 1995) and that the reader should take heed of unavoidable impurities and always stick to the Original.

A banana allegory might well serve here as a way to change our perspective on the problem of translations, and an invitation to reconsider the various fates of original and copies: *La grande bananeraie culturelle* (1969–1970), a work by Gérard Titus-Carmel, juxtaposed 59 plastic bananas and one natural banana, the original model and the 59 copies. During the exhibition, the viewer was confronted with their contrasting fates: while the original, natural almost parental Form was inexorably decaying, the 59 copies or offspring, more or less indiscernible or anonymous, were still glowing in all their artificial splendour (Derrida 1978: 285)[3]. Titus-Carmel's work is an invitation to reassess the relationship between the Original and the copies, and to focus on the perplexing glow of the latter.

Another linguistic anecdote suggests the dignity of translations. In the French comic book *Astérix le Gaulois*, Obélix's little white dog is named Idéfix. Having to render the original pun on the French expression *idée fixe* (fixed idea or obsession), the translators came up with Dogmatix. The English name, no doubt, is even better than the French original, since it keeps the meaning of the original pun, and even adds to it, by reference to the dogness of the character and the suggestion of dogmatism. I use the phrase "*Dogmatix miracle*" to label this category of overreaching translations, which keep all the assets of the original word or phrase, and even sur-

---

3   In a Derridean framework, one could claim that translations are like a *supplement:* a concept which suggests both addition to and substitution of the Original (see Derrida 1967: 203). On Original and copies, see also Hergé's *The broken ear* (*L'oreille cassée*) (1964/1993) and the analysis given by Clément Rosset (1977: 146–153).

pass it. Maybe the Dogmatix miracle is a category of its own, and not a single occurrence in the history of translations. A surprising example of the Dogmatix miracle can probably be found in Norman Kemp Smith's translation of Kant's *Kritik der reinen Vernunft*, described as "*a classic of philosophical translation*". Originally issued in 1929, it was recently reprinted with an introduction by Howard Caygill that states:

> "Kemp Smith's version of the Critique has become the translation of reference, serving as the touchstone of quality not only for subsequent attempts at translation, **but also, apocryphally, for the original German**." (Kant 2007: v; emphasis added).

The case of the Kemp Smith's translation might well be urban legend, popular in anglophone philosophy departments; the aim of this paper is not to verify its claims. My goal here is to show that translations are not only covering the original meaning or relocating it in different contexts. I want to show how the Original itself is constantly challenged in its legitimacy by the multiplication of its copies; and how the copies constantly renegotiate the initial understanding that one could have of the Original. Whether or not they pertain to the Dogmatix miracle category, translations have something to teach us about the Original. This introductory fable on dogs and bananas will be tested on the case of Darwin's phrase *natural selection*.

THE LOSS OF AN "S": IMPORTING NATURAL SELECTION IN FRANCE

In France, translating the *Origin* was no easy job and Clémence Royer (1830–1902), Darwin's first translator, has been the target of many criticisms. Above all, she has been accused of having loaded Darwin's text, intentionally or not, with Lamarckian insights and the introduction of designs or intellectual powers absent from the original. Historians may point at (intentional?) blunders such as Darwin's "*power always intently watching*" translated as "*un pouvoir intelligent (…) constamment à l'affût*" (Darwin 1859: 189; Conry 1974: 263).

While translating a technical text like the *Origin*, a translator has two different possibilities: *to transliterate* or *to translate*. In the *Magasin pittoresque*, an anonymous review entitled "*Sélection naturelle. Choix de la nature*", published in September 1860, both *transliterates* natural selection into "*sélection naturelle*", and *translates* it as "*choix de la nature*" (Anonymous 1860). As the term was new and unknown to the public, it was immediately explained as a way of choosing, or sorting out:

> "C'est ce que M. Darwin appelle la sélection naturelle, le choix ou le triage qu'amènent les circonstances, que transmet le principe d'hérédité et qu'entretient la lutte incessante engagée entre tous les êtres organiques…" (Anonymous 1860: 295).

Translating the English term into French requires an attempt at interpreting the meaning of the original and at finding a possible equivalent. Thus, Clémence Royer (1862) rendered "*struggle for life*" as "*concurrence vitale*" (literally: "*vital competition*"). As to natural selection, Royer wound up using the equivalent phrase "*élection naturelle*". The question immediately arises: Why "*élection*" and not "*sélection"? Is Clémence Royer to blame? In other words, was it possible, in 1862, to translate "*natural selection*" in French by "*sélection naturelle*"?

Selection might have been easily transliterated in French. But the issue is different with the verb "*to select*" and the derived adjective form "*selected*". *Election* and *selection* form in English two etymological twins or a doublet, directly modelled on the Latin: the verb *eligere (electus)* means to tear or dig out, and *seligere (selectus)* means to choose and put together, as is explained in any etymological dictionary (see Table 1). But, strikingly enough, if the term *eligo* was directly translated into the very common French verb *élire*, there was no direct equivalent for *seligo* and the derived adjective *selectus* was traditionally translated as "*choisi*". In French, there was no verb available to express "*to select*". Accordingly, the easy way was to render "*to select*" by the French "*choisir*". This solution was retained by later translators of the *Origin*, like Jean-Jacques Moulinié or Edmond Barbier[4]. But it entails the following problem: when a French reader will read "choisir" in Darwin's text, he/she will not immediately make the mental connexion with the operation of "selection". Besides, *choisir* (to choose) is a very common verb, and it loses the technical aspect of "selection". This is why Moulinié occasionally recurs to neologisms: for instance, in the *Introduction* of the *Origin*, he translates Darwin's phrase "*naturally selected*" into "*naturellement conservé ou sélecté*", but in the following sentence, "selected variety" becomes "*variété ainsi épargnée*" (Darwin 1873: 4). As to Barbier, he avoids the verbal and adjective forms as much as he can. He translates the same passages with "*être l'objet d'une sélection naturelle*" and "*variété objet de la sélection*" (Darwin 1876a: 4).

Table 1. The Latin doublet *eligo/seligo*.

Latin	English	French
Eligo, electus	To elect, elected	Élire, élu
Seligo, selectus	To select, selected	– Choisir, choisi? – Sélire? – Séliger? – Sélectionner?

The Geneva psychologist Edouard Claparède (1832–1871) suggested that "*élection*" and the verb "*élire*" were an easy way to stay close to the English terms *selection, to select* and *selected*, without creating neologisms (Claparède 1861: 534). Claparède's initial suggestion was followed by Clémence Royer in the first edition of the French *Origine* (Darwin 1862).

Given the absence of any exact equivalent for *selection / to select*, the *élection/ élire* solution indeed appears to be a very elegant translation[5]. "*Élection*" might sound strange in the context of natural processes, but precisely *because* of that reason, it might convey a sense of the technical character of the English term. Most of

4   See, for instance, Darwin (1873: 31), where "*selected for breeding*" was translated as "*choisi pour la reproduction*". See also Darwin (1876a: 31).
5   For a defense of Royer as a translator, see Miles (1989).

all, it provides an equivalent family of words for the derived forms of *to select* (*selective*, *selected*). I claim that this lexical problem is the main reason why Royer used "*élection*".

Other attempts were made but all proved unsatisfactory. Some suggested that the Latin verb *seligere* could just be transcribed in French as "*séliger*"[6]. Royer herself suggested that she could have used the neologism *sélire*; it would have sounded elegant but would also have been perplexing. The worse, or so claims Royer, would probably be the ugly and badly formed "*sélectionner*"[7].

The problem with the term "*élection*" is its connotation that choice is *personal*. "*Natural selection*" was often caricatured in France as an oxymoron or a *contradictio in adjecto*. The most representative character of this tendency is surely Pierre Flourens (1794–1867), the perpetual secretary of the Académie des Sciences in Paris. A major thrust of Flourens' critique of the *Origin* bears on the use of figurative language, and on the danger of "*personifying Nature*" (Flourens 1864: 2) – a fallacy very common in the eighteenth century but which Cuvier had supposedly eradicated in the nineteenth.

Royer's French translation of "*selection*" as "*élection*", can partially account for Flourens' belief, that the term "*élection naturelle*" suggests that nature has "*a power to elect* [un pouvoir d'élire], *a power comparable to Man's*" (Flourens 1864: 6). For Flourens, *élection*, whatever it may be, implies the consideration of an intellect, of some intellectual power, which *chooses*. *Élection* inescapably means "*to choose consciously*" and Flourens sharply criticized such a phrase as "*élection inconsciente*" [unconscious selection], which Darwin used in the first chapter when he described the practice of breeders (see for instance Darwin 1859: 36):

> "Either natural élection is nothing, or it is nature; but nature gifted with élection, but nature personified: last error of the last century. The Nineteenth Century does not make personifications any longer." (Flourens 1864: 53).

For Flourens, the case was conclusive: Darwin was falling back into "*gibberish*" [galimatias], "*pretentious and empty jargon*" [langage prétentieux et vide], "*childish and outmoded personifications*" [personnifications puériles et surannées], all things blatantly contradicting the "*sturdiness of the French spirit*" [solidité de l'esprit français] (Flourens 1864: 65). Weirdly enough, Flourens personified the

---

6   This occurs in an attempt to render some very specific nuances of the German debate. A French translation of Ludwig Büchner reads: "*Dans la pensée de Darwin, la nature n'amende pas* [züchtet nicht] *comme l'homme peut faire, simplement, elle élimine, elle sélige* [wählt aus], *mais sans parti ni dessein.*" (Büchner 1869: 27).

7   "*Sélire*" would have been a better translation of "*to select*" than "*sélectionner*". In fact, "*sélectionner*" is based on "*sélection/selectio*", which is itself derived from the verb *seligere*. Thus, the new verb "*sélectionner*" is at best redundant, at worst, a lexical monstrosity. Unfortunately, the evolution of French confirmed Royer's worst nightmares, since *sélectionner* made it into common language, being now a part of the ordinary lexicon of football players, who, not unlike cattle, cats, dogs or race horses also have their "*sélectionneur*". We have a similar case in contemporary French with the term "*solution*", which comes from the Latin "*solvere*"; the verb associated with it is "*résoudre*", but people start to use the most ugly *solutionner*, based on the substantive *solution*.

"*Nineteenth Century*" while claiming that the time for personification was over! As a reward, he himself ended up being considered as mere gibberish (as indicated by Conry 1974: 30).

Flourens blamed Darwin for giving up Cuvier's "*école des faits*" [the school of facts] just as Adam Sedgwick chastised him for having "*deserted—after a start in that tram-road of all solid physical truth—the true method of induction*" (Sedgwick to Darwin, 24 November 1859; Darwin 1991: 396). Darwin was guilty of *esprit de système* and of having abandoned sound observation. In the strong words of Bishop Wilberforce (1860: 250), "*In the name of all true philosophy we protest equally against such a mode of dealing with nature, as utterly dishonourable to all natural science, as reducing it from its present lofty level as one of the noblest trainers of man's intellect and instructors of his mind, to being a mere idle play of the fancy, without the basis of fact or the discipline of observation.*" The physiologist Claude Bernard in his *Introduction to experimental Medicine* (1865/1984: 140) shared the same position and ranked Darwin among Romantic German *Naturphilosophie*.

One cannot help but wonder about the effects of dropping a single letter! If Royer had not rendered the English *selection* by the French *élection*, would the fate of Darwinian theory in France have been any different? It is striking that, when Royer finally had to give up the easy "*élire*" in the second edition of her translation (1866), she justified her previous choice of "*election*" on the basis of strictly linguistic constraints:

> "As to the term selection, seeing that it has been adopted by most of Mr. Darwin's critics, and that these competent naturalists have not stepped back in front of this neologism—one that had seemed useless to me—, I finally decided, though reluctantly—to use it. I had to take upon myself to introduce in French the adjectives sélectif and sélective, which I could not avoid [...]. In giving up the word élection, that I had used in my first edition, I have compromised with the opinion of the great number, but I have to confess that my conscience is not at peace with this sacrifice." (Royer 1866: xii–xiii).

For Royer, Flourens was wrong in suggesting that "*élection*" presupposes an intellectual power. Nobody objects to the chemists' *elective* affinities, which are merely blind natural forces (Darwin 1866: xiii). The issue of translating "*natural selection*" casts an interesting light on the ways one should talk about nature and about the necessity of resorting to metaphorical language in science.

THE CULTURE OF FRENCH BREEDERS

The absence of the term "*sélection*" in French might be surprising and must be further documented. In 1859, "*selection*" had been a very common English term for a while already (for the history of the term in English, see the *Oxford English Dictionary* and Hodge 1992). In the special meaning of "*action of a breeder in selecting individuals from which to breed, in order to obtain some desired quality or characteristic in the descendants*", the term can be found in John Sebright's 1809 pamphlet, where one can read: "*Were I to define what is called the art of breeding, I should say, that it consisted in the selection of males and females, intended to breed together, in reference to each other's merits and defects*" (Sebright 1809: 5). Or, according to William Youatt's famous statement, that Darwin quotes in the *Origin* (1859: 31):

> "These causes, however, would operate only to a limited extent; a more powerful principle would, at a very early period of sheep-husbandry, be called into action—that which enables the agriculturist not only to modify the character of his flock, but to change it altogether—the magician's wand, by means of which he may summon into life whatever form and mould he pleases—the principle of selection—the fact, that 'like will produce like'." (Youatt 1840: 60)

In French, however, *sélection* (in the sense of choice) was very rare. The *Grand Robert de la Langue française* dates initial usage of the word in 1801 but remarks it is directly imported from English. Everytime we have encountered the term "*sélection*" in a French treatise on husbandry or in a periodical for agriculture, it was always cursorily and in relation to the British context. For instance, Lefebvre-Sainte-Marie refers cursorily to improving animal breeds "by crossing or selection in the breed itself" (1849: 219). Similarly, Eugène Gayot in his *France Chevaline* evokes the fact that "the success of consanguineous alliances bears on a well-understood *sélection*" (1850 (part 2, vol. 3): 26). Everytime French breeders refer to selection, they think of it in relationship with crossings. They are far from imagining the patient step-by-step accumulation that is so characteristic of Darwinian selection.

It might be surprising that France did not have an equivalent term for "*selection*" since it had a long tradition of royal stud farms. But it seems that the practice of choosing the characteristics one wishes to transmit followed by selecting the genitors was not theorized as such. By contrast, the fate of *natural selection* in French demonstrates the singularity of British breeding practices. The absence of a well-established term for *selection* sheds light on the existence of two different breeding cultures in Britain and on the Continent.

On the lexical level, I will give two examples: John Sinclair's *Code of Agriculture* (1821) and William Youatt's *History of the Horse* (1834). First, let us compare two versions of John Sinclair's *Code of Agriculture*: the 1821 third English edition, and the 1824 French translation by Matthieu de Dombasle (Sinclair 1821; 1824). Sinclair (1754–1835) was a Scottish politician, with a strong involvement in agriculture. His *Code* is one of the books that made the works of John Sebright and Robert Bakewell known to the French public. In the French version, the word selection was systematically erased. If we turn to the section "*On the origin of improved breeding*", we read:

> "The art of improved breeding consists, in making a careful selection of males and females, for the purpose of producing a stock, with fewer defects, and with greater properties than their parents; by which their mutual perfections shall be preserved, and their mutual faults corrected." (Sinclair 1821: 104).

But the French translation of the section ("*Des principes de l'amélioration des races*") reads:

> "L'art d'améliorer les races de bestiaux, consiste à faire **un choix judicieux** des mâles et des femelles, employés à la reproduction, afin de produire une race qui ait moins de défauts et de

meilleures qualités que les races originaires, en conservant les perfections de ces races et en corrigeant leurs défauts mutuels." (Sinclair 1824: 186–8; emphasis added).

And again when the original English text refers to Bakewell's achievements:

"It was upon this principle of selection, that Bakewell formed his celebrated stock of sheep, having spared no pains or expense, in obtaining the choicest individuals, from all the best kinds of long or combing woolled sheep, wherever they were to be met with."

But the French version simply states: "*C'est* sur ce principe *que Bakewell, etc.*" (Sinclair 1824: 188). It is striking that the word selection is either simply erased or translated by a word of the family of choice (*choisir*).

A similar case can be documented in the translation of William Youatt's *History of the Horse* (1834)[8]. When the English version goes:

"A horse with a shoulder thicker, lower, and less slanting, than would be chosen in a hackney, will better suit the collar; and collar-work will be chiefly required of him. A stout compact horse should be **selected**, yet not a heavy cloddy one." And below: "has **selected** one with sound feet" (Youatt, 1834: 26; emphasis added)

The French translation simply reads:

"Afin de mieux tirer au collier, ses épaules doivent être plus fortes, plus basses et plus obliques que celles du cheval du selle. Comme il est principalement destiné au service du collier, on doit le **choisir** vigoureux et ramassé, sans être lourd." And a few lines below: "s'il l'a choisi avec le pied sûr." (Youatt 1851: 229; emphasis added).

Similarly, when Youatt writes (1834: 31): "*Surely the breeder might obviate this. Let a dray mare be **selected** as perfect as can be obtained.*", the French translation goes (1851: 250–1): "*Assurément l'éleveur peut y porter remède. Qu'une jument de trait aussi parfaite que l'on pourra l'obtenir, soit **choisie**.*"

Those two examples confirm that the terms "*selection*", or "*to select*" were not available in French before the introduction of the Darwinian theory. They could occasionally be found in French texts on breeding, but they were always used as *hapax*, with reference to the British practices.

One cannot evoke breeding cultures in France in the 1850s without reference to the work of the Vilmorin's family, which was accurately described by Gayon & Zallen (1998). Louis de Vilmorin (1816–1860) reissued, shortly before his death, a collection of previously published papers (Vilmorin 1859). This collection includes a paper by his father Philippe-André de Vilmorin (1776–1862), his "*Notice on the improvement on the wild carrot*" (read before the Horticultural Society of London, on March 3, 1840). According to this paper, the question of the origin of domesticated plants bears on the transformation of some weak and filamentous substances into juicy and bulky roots, in edible plants. The "*means by which this has been effected*" are mostly unknown, since most of the vegetables have been simply trans-

---

8   I have compared the text of Youatt (1834), available online at <http://www.archive.org/stream/historyofhorsein00youa/historyofhorsein00youa_djvu.txt>, with the French version at the Bibliothèque Nationale de France in Paris (Youatt 1851). The French edition does not follow the order of the original text, which renders difficult the search for corresponding passages in the two versions.

mitted to us, once they were "*all-shaped*" [tout-façonnés] (Vilmorin 1859: 6–7).
Garden plants constantly vary or, in Vilmorin's words, they tend to be "*loose*" [ten-
dent sans cesse à jouer] (1859: 7). This means ordinarily that they degenerate, or
return to their primordial form. This is very important to him: modifications have
been effected in the ancient times, and the secret is mostly lost to us. For Vilmorin,
some means to achieve the transformation of recent wild forms are still available,
for instance to nourish the plant in an over-abundant fashion, what he calls to treat
the plant "*gardenly*" [traiter jardinièrement] (1859: 8). But he thinks that this means
is clearly insufficient to achieve sustainable results. When he describes his experiments with wild carrots, repeated year after year or generation after generation,
words like "*to select*" or "*to pick*" are conspicuously absent (1859: 9–14). Vilmorin
rather describes his achievement as "*a kind of creation*", but not a "*real conquest*",
since we already had the carrot (1859: 14).

As to Louis de Vilmorin, he has worked on beetroot with approximately the
same results. His problem was the following: how the various variations that occurred were "*fixed by perseverance and care, taken in choosing the individuals to
breed*" (1859: 15). "*Choice*" was Vilmorin's word, be it "*constant*" [on choisisse
constamment] (1859: 16) or "*scrupulous*" [choix attentif des individus reproduc-
teurs] (1859: 18)[9].

Vilmorin's works in thornless gorse (*Ulex europaeus*), published in 1851, is
also very interesting on the issue of breeders' practices. Vilmorin described his
problem as "*fixing, in a sustainable manner, a modification that has only been so far
a temporary monstrosity*" (1859: 31), or as "*protecting*" the variations in order to
"*fix*" them (1859: 35). Vilmorin's paper described an interesting method which he
called "*maddening the plant*" [affoler la plante] (1859: 36), [l'affolement] (1859:
37). By this, he meant the art of choosing any deviation, not the ones that are the
closest to the target form; but the form that is the most remote from the original
plant. Vilmorin did not aim at selecting a definite trait by accumulating minute
variations, but at choosing individuals, which showed the greatest propensity to
vary and deviate[10]. In this process of "*maddening the plant*", hybridization might
play a role, as it increases the tendency to vary.

Differences in terminology reveal that the breeding cultures were very different
in England and on the Continent, and therefore, complaints about the poor results
of French breeding have to be taken seriously. For instance, when Prosper Lucas, in
his major treatise on hereditary diseases (1847a–b)[11], gave high praise to the incredible results of English breeders, he wanted to explain to his French readers the
principles of the English method of *breeding in-and-in*. Lucas apparently under-

9   Gayon & Zallen (1998: 246, note 8) suggested: "*On rare occasions (as in the 1851 paper)
      Vilmorin employed the term 'selection'. Most typically however, he referred to this process as
      'choice'.*" I have not been able to find the word "*selection*" in the French version of the 1851
      paper. In any case, if the term should occur, it was, to say the least, extremely rare.
10   This crucial difference seems to have been overlooked by Gayon & Zallen (1998).
11   Lucas' treatise was quoted by Darwin (1859: 12) as "*the fullest and the best on this subject*"
      (i.e. on "*inheritable deviations of structure*").

stood the issue of breeding as a problem of heredity, rather than selection[12]. But he lacked the words to give a proper description: he wound up saying that it consists in "*specifying* [préciser] *the character wished for*", and then "*making election*" [faire élection] of the proper genitors, males and females which elicit this character (Lucas 1847a: 203). Interestingly enough, Lucas used *élection* obviously as an equivalent of *selection*. Not that the term, its presence or absence, accounts for the results of breeders, in France or England; but the lack vs. existence of accurate terms is a quite revealing symptom of the state of practices. This brings further evidence for Royer's case: when translating *selection* by *élection*, she was neither very original nor guilty of the crime of Lamarckianising Darwin. She was just filling in the blanks of the French technical vocabulary and elaborating on a sound and efficient lexical possibility.

GERMAN WANDERINGS AND WAVERINGS

The diffusion of Darwin's *Origin* in the German context was carefully analysed by Sander Gliboff (2008). During Darwin's long delay, Gliboff showed, the researches led by the first German translator Heinrich Georg Bronn (1800–1862) paralleled Darwin's in some uncanny ways. Darwin was one of Bronn's authorities on biogeography and uniformitarian geology. Conversely, Bronn's books were among the most heavily annotated works in Darwin's library. Therefore, Gliboff resisted the easy conclusion that "*the Germans mangled or misconstrued Darwin's obvious meaning*" and he convincingly argued that "*the process of bringing Darwin's* Origin *to Germany and making it understood there clearly involved much more than a mere mechanical substitution of German words for English in a text*" (Gliboff 2008: 4).

In this section, I will use the German language as a means to further analysing the phrase "*natural selection*" and diffracting its various meanings. Several translations have been proposed to render this mysterious English phrase (see Table 2). Bronn initially attempted to find a German phrase that would accurately render Darwin's thinking. But after a few years, German biologists simply gave up and decided to *transliterate* rather than translate *natural selection* into "*natürliche Selektion*".

---

12   See for instance, Lucas (1847a: 205): "*Such results should strongly encourage the English industry to persevere in applying* the principle of the heredity of volume *to every species proper to feeding*" (emphasis added). It seems that the word "*heredity*" was rare in English in its biological sense. Darwin (1859) did not refer to "*heredity*" but to "*inheritance*".

Table 2. German translations of *selection*.

Reference	Translation of the English term *selection*
Bronn 1860*a*	*Wahl der Lebensweise*
Bronn 1860*b*; Weismann 1893	*Züchtung*
Seidlitz 1871	*Auslese*
Carus *in* Darwin 1876*b*	*Zuchtwahl*
Bronn 1860*b*, Büchner 1869	*Auswahl*
Weismann 1886, 1909; Plate 1903	*Selektion*

One obvious conclusion is that natural selection is simply impossible to translate. But the various attempts of the German translators help us disentangle the various tensions embedded in Darwin's phrase. "*Natural selection*" connects different features that are difficult to find within a single German equivalent: "*Züchtung*" refers to the practices of breeders, whereas "*Wahl*" means choice; "*Auswahl*" indicates the act of sorting out, whereas "*Auslese*" suggests both choice and elimination. These nuances might seem minimal but each one emphasizes a different facet of the Darwinian concept.

Bronn (1860a) first came up with the phrase "*Wahl der Lebensweise*" [choice of the way of life]. As he understood it, natural selection describes the fact that some of the offspring may diverge from the original stock as a result of choosing different nutritional supplies and altogether "*another way of life*". A different use of the organs results from this choice that in turn entails a diverging process, both from the parental form and the unchanged siblings.

Although the emphasis on "*Lebensweise*" could seem pretty close to what is called "*line of life*" in the *Origin* (Darwin 1859: 321, 339), Darwin harshly rebuked Bronn's suggestion, saying that "*It leaves the impression on my mind of the Lamarckian doctrine (which I reject) of habits of life being all important*" (Darwin to Bronn, February 14, 1860; Darwin 1993: 83). Correlatively, Bronn was encouraged to try harder and find a better equivalent. To assist his translator, Darwin gave Bronn two different reasons for using the term "*natural selection*": before all, "*its meaning is* not *obvious & each man could not put on it his own interpretation*"; but equally important is the fact that the phrase, "*at once connects variation under domestication & nature*". Natural selection depends on the analogy between domestic and wild entities. Darwin accordingly thinks Bronn should look for "*any analogous term used by German Breeders of Animals*". He goes as far as to suggest that *Adelung*, ennobling, may be a good choice, although "*perhaps too metaphorical.*" (Darwin to Bronn, February 14, 1860; Darwin 1993: 83). The analogy was so central to Darwin's thinking that he seemed ready to accept any term suggesting that same rationale, even one heavy with progressionist connotations, like *Adelung*[13].

---

13   On "*Adelung*", Sander Gliboff (2008: 136) suggested that Darwin probably meant *Veredelung*, which refers to grafting. Dictionnaries do not have any agricultural usage for *Adelung*.

When Bronn introduced the term *natürliche Züchtung* as a translation for *natural selection*, he carefully transcribed Darwin's reasons in a footnote: "*the constantly recurring term* Selection" is "*not of common use in England*" (Darwin 1860a: 10). Bronn added that "*Auswahl zur Züchtung*" [choice for breeding]. He also suggested that "*Auswahl zur Nachzucht*" [choice for subsequent breeding] would probably have been better than "*Züchtung*", and by this term, Bronn meant nothing more than "*the sorting out of the domestic animals*" [die Auswahl der Zucht-Thiere]. Bronn also considered the idea that the neologism "*Zuchtwahl*" [choice for breeding] might be better, especially in regard to the phrase "*sexual selection*" (Darwin 1860a: 10). "*Wahl*" implies that an agent makes a conscious choice, and this is probably why Bronn suggested it applies to sexual, rather than natural, selection. Bronn again referred to terminological issues on page 87, when introducing the definition of the term *natural selection* (in Bronn's terms: "*Natürliche Auswahl oder Züchtung*") as "*a principle of preservation*". As Sander Gliboff (2008: 137) aptly remarked: "*when Darwin wrote about* artificial, natural *and* sexual selection, *it sounded consistently like three versions of the same process, but it was not so in Bronn's German*".

Darwin's subsequent translator Victor Carus finally adopted the term *Zuchtwahl*, which seems most efficient at combining both breeding and choice. But other words focusing on the choice or eliminative dimension of selection will be put forward, such as Seidlitz's *Auslese* or Büchner's *Auswahl*. The case of German shows that natural selection nicely combines different sets of meanings: breeding and choice. Darwin made clear in his correspondence with Bronn that he had two different agendas in using the term natural selection: first, to use a rare term that would force his readers into careful attention; second, to constantly suggest the powerful analogy of nature's processes with domestic productions.

A HOST OF FORERUNNERS?

After the publication of the *Origin*, Darwin immediately received a lot of correspondence, from readers who claimed that they had forestalled him. Darwin's letter to Baden Powell (18 Jan 1860) is very helpful to disentangle that, which Darwin "originated" from that, which he actually borrowed—but from whom?

"My health was so poor, whilst I wrote the Book, that I was unwilling to add in the least to my labour; therefore I attempted no history of the subject; nor do I think that I was bound to do so. I just alluded indeed to the Vestiges & I am now heartily sorry I did so. No educated person, not even the most ignorant, could suppose that I meant to arrogate to myself the origination of the doctrine that species had not been independently created. The only novelty in my work is the attempt to explain how species become modified, & to a certain extent how the theory of descent explains certain large classes of facts; & in these respects I received no assistance from my predecessors. To the best of my belief I have acknowledged with pleasure all the chief facts & generalisations which I have borrowed. If I have taken anything from you, I assure you it has been unconsciously; but I will reread your Essay. Had I alluded to those authors who have maintained, with more or less ability, that species have not been separately created, I should have felt myself bound to have given some account of all; namely, passing over the ancients,

Buffon (?), Lamarck (by the way his erroneous views were curiously anticipated by my Grand-
father), Geoffroy St. Hilaire & especially his son Isidore; Naudin; Keyserling; an American
(name this minute forgotten); the Vestiges of Creation; I believe some Germans. Herbert Spencer; & yourself." (Darwin 1993: 39).

In response to this kind of letters, Darwin added a "*Historical sketch*" to the third
edition of the *Origin* (Johnson 2007), where he pays due tribute to previous contributors to the theory of descent with modification or the discovery of natural selection. Considering these various contributions, one has to distinguish those who
may perhaps have a *concept* of natural selection but certainly don't have the *phrase*,
and those who have the *phrase* but don't know exactly what to do with it. Between
the phrase and the concept, is there really a difference, the historian of science asks.
Can one have a concept if one doesn't have the *term* to name it? To put the question
in Daryn Lehoux's funny way (2006: 541): "*If the history of the Frisbee cannot be
pushed any farther back than the changing of its name from Pluto Platter, then we
are only talking about the history of a word, not of a thing*"—and certainly not of a
concept, one might add. However, I claim that what may apply to the Frisbee may
prove unsatisfactory in the case of natural selection, because the term "*natural selection*" does really matter and is a crucial part of the concept itself. Therefore, the
challenge of translating "*natural selection*" into different languages is really central. Can one have a concept of natural selection in languages others than English?
Is *natural selection* special in any sense, compared with other scientific concepts,
such as *attraction*?

   The question becomes critical when considering who has discovered natural
selection, or who was the closest to forestall Darwin's discovery (applying natural
selection to account for evolution). Conway Zirkle (1941) has collected many instances of *natural selection* before Darwin, portraying a colourful gallery of forerunners, anticipators or quasi forestallers of the Down House naturalist. But Zirkle's abundant harvest is not very convincing and it shows more Zirkle's complete
unawareness of the importance of terms, than actual occurrences of *natural selection*. Zirkle clearly documents that the word "*selection*" was used in reference to
human activity, be it directed on non-human or human animals (when marriage is
concerned). In many cases, Zirkle is victim of taking vague resemblances for true
similarities.

   The case of Charles Victor Naudin (1815–1899) is a decisive test-case regarding my argument that the term "*natural selection*" was extremely rare in French, if
it existed at all, when Darwin's ideas were first introduced in France. According to
Zirkle (1941: 121), "*Naudin in the* Revue horticole *(pp. 103–105) came even closer*
[than Spencer] *to forestalling Darwin. Obviously so many biologists were ap-
proaching the conception of natural selection that it was only a question of time
until it would break out into the open*". Some of Darwin's contemporaries, like the
Belgian-born botanist Joseph Decaisne (1807–1882), fell into the same mistake.
Darwin complained in a letter to Charles Lyell, on December 22, 1859: "*I have not
seen Naudin's paper & shall not be able till I hunt the Libraries; I am very curious
to see it. Decaisne seems to think he gives my whole theory.*" Naudin very clearly
paralleled Nature's procedures with the way we proceed in creating our varieties.

He often suggested that *"it is her very procedure that we have transported to our practice"* (1852: 104). However, accordingly to what we have claimed above, Naudin never used the word *selection*, neither to describe human procedures, nor to describe Nature's. As Darwin wrote to J.D. Hooker, when sending him back his copy of Naudin (23 December 1859):

> "I am surprised that Decaisne shd say it was same as mine. Naudin gives artificial selection as well as a score of English writers; & when he says species were formed in same manner I thought the paper would certainly prove exactly the same as mine. But I cannot find one word like the Struggle for existence & Natural Selection. On the contrary he brings in his principle (p. 103) of Finality (which I do not understand) which he says with some authors is fatality, with others Providence, & which adapts the forms of every Being, & harmonises them all throughout nature." (Darwin 1991: 444).

Darwin's argument can be read on different levels. First, he emphasizes the linguistic aspects of the question: the absence of words like *struggle for existence* or *natural selection* is really a problem. Besides, there are strong divergences in their philosophy of nature, Naudin referring to *finality* as a sort of metaphysical principle of harmonisation. This is why Darwin considers that he *"cannot see much closer approach to Wallace & me in Naudin than in Lamarck—we all agree in modification & descent"* (23 December 1859; Darwin 1991: 444, emphasis in original).

"*Wallace & me*" is a very strong way to mark the unity between the two English naturalists, and their difference with others (such as Lamarck and Naudin). But if Darwin's statement of his unity with Wallace relies on the sharing of words, then he probably means *"struggle for existence"* rather than *"natural selection"*. In fact, Darwin seems to have neglected, at times, the importance of using the phrase *natural selection*. Had he taken it more seriously, he might not have been so shocked upon receiving Wallace's Ternate manuscript in March 1858 and certainly would not have said: *"I never saw a more striking coincidence. If Wallace had my M.S. sketch written out in 1842 he could not have made a better short abstract!"* (Darwin to Lyell, 18 June 1858; Darwin 1991: 107).

And for one good reason: while Darwin and Wallace's contributions to the famous joint reading and publication at the Linnaean Society in 1858 might appear close in spirit, they nonetheless differ in one crucial regard, namely that Wallace never refers to *natural selection*[14]. Wallace had read Lyell and Malthus and he referred to *"the life of wild animals"* as *"a struggle for existence"* (1858: 54). Wallace's Ternate paper includes such reasoning as:

> "An antelope with shorter or weaker legs must necessarily suffer more from the attacks of the feline carnivora; the passenger pigeon with less powerful wings would sooner or later be affected in its powers of procuring a regular supply of food; and in both cases the result must necessarily be a diminution of the population of the modified species. If, on the other hand, any species should produce a variety having slightly increased powers of preserving existence, that variety must inevitably in time acquire a superiority in numbers." (Wallace 1858: 58).

---

14   For recent attempts to distinguish between Wallace and Darwin's theories, see, e.g., Bock (2009) and Gayon (2009).

Such arguments probably sustain Wallace's claims as the co-discoverer of natural selection. For many readers of the joint papers, Darwin's "*natural selection*" simply *generalises* what Wallace has vividly *illustrated* (see for instance Owen 1860: 509). But although he had an idea of species changing due to what may be called population pressure, he never used the phrase "*natural selection*" to designate it.

In spite of these terminological divergences, Wallace struggled hard to show retrospectively that he did not come across the idea by mere chance and that, while he lacked the term, he had the concept. In particular, he devoted a great deal of energy to contrast his own case with the ones of Patrick Matthew and William Charles Wells. Matthew and Wells had certainly propounded the fundamental principle of natural selection but they had made no further use of it and had failed to see its applications; on the contrary, Wallace made clear that he "*both saw at the time* [1858] *the value and scope of the law which* [he] *had discovered, and ha*[s] *since been able to apply it to some purpose in a few original lines of investigation*." (Wallace 1870: iv). The contrast between Wallace and Matthew is especially interesting since the latter came very close to using the phrase. The *Historical sketch* added to the third edition of the *Origin*, helps us but poorly to understand how Darwin publicly analysed their respective cases. Darwin (1959: 62–63) acknowledged that Matthew "*clearly saw the full force of the principle of natural selection*", whereas Wallace "*arrived at almost the same general conclusion that I have*" (Darwin 1959: 71). We know that Darwin felt very close to Wallace, and his supporters suggest that the concept might be present even if the term is absent, although the concept cannot be present if some systematic application of the principle is not made. But what about the far less known case of Matthew?

As Matthew himself proclaimed in 1860, his 1831 essay *On naval Timber and Arboriculture* presents "*nature's law of selection*" (Matthew 1860). The extract published in 1860 refers twice to selection: one is only to artificial selection, Matthew referring to man "*preventing deterioration, by careful selection of the largest or most valuable as breeders*"; the other is about "*the infirmity of* [the] *condition*" of varieties produced by man, "*not having undergone selection by the law of nature*". The juxtaposition of the theme, of the word "*selection*", and the expression "*law of nature*", is very akin to Darwin's actual phrase "*natural selection*". In another passage (1831: 308), Matthew refers to "*man's interference, by preventing this natural process of selection among plants*". Whereas Matthew fully acknowledged that Darwin "*has more merit to the discovery*", he also claimed that he had identified not only *selection*, but selection as "*a law universal in Nature, tending to render every reproductive being the best possibly suited to its condition*" (Matthew 1860: 312).

In contrast to Matthew's emphasis on the term *selection*, which is the basis for his claims, it is striking that Wallace never really accepted the term "*natural selection*". Correlatively, he saw no direct link between his ideas and selection in domestic animals, whereas, by contrast, breeders' practices provided Darwin with a powerful analogy. Clear evidence of this point is provided by the fact that Wallace crossed out "*natural selection*" in his copy of the *Origin* and that he substituted "*survival of the fittest*" for it (see Beddall 1988: 275, and Browne 2002 for a picture).

This terminological take on the question of the forerunners sheds light on the question: what did Darwin *invent?* Surely not the term "*selection*", nor the idea of "*modification of species*", but it seems that he had a peculiar theory of evolution, which is characterised by a constant use of *natural selection*, be it a phrase, a process, a mechanism, and maybe all of them. In this case, who can be said to be the closest to Darwin's "*invention*" or "*discovery*" of *natural selection*: the forerunner with the term but no further application (Matthew); or the co-discoverer with the application but without the term (Wallace)? I claim that the technical term *natural selection* matters if one wants to understand Darwin's mechanism and the analogy with the practice of English breeders on which it rests. Of course, Darwin always felt closer to Wallace than to Matthew. But for failing to fully accept the term, Wallace was led to several misunderstandings. Neither the concept, nor the word, were new, but Darwin's articulation of the two was his own brilliant innovation. Darwin invented the hypothesis that nature selects, just as breeders do, and that this natural selection is the main explanation of the diversity we find in the living world.

REFLECTING BACK ON THE ORIGINAL

I would like to show now that the translators' difficulties are not simply variations on the well-worn theme of "*traductor-traditor*". I claim that the issue of translations also helps us understand some difficulties in the English wording of Darwin's text. As Gillian Beer aptly noted, in the course of the successive editions of the *Origin*, "*long paragraphs agglomerate around the terms 'natural selection' and 'nature', paradoxically in an attempt to cut back their superplus of meaning*" (Beer 1996: xxiv).

The history of the phrase "*natural selection*" in Darwin's texts has been thoroughly described by Jonathan Hodge (1992), whom I follow here (see also Hodge & Kohn 1985). The phrase is absent from Darwin's early *Transmutation Notebooks*, where only expressions like "*picking*" or "*sort out*" occur[15]. Darwin has been using the word "*selection*", routinely by 1839 but the earliest known use of the actual phrase "*natural selection*" is in a manuscript text of 1841, "*where its use is not marked by any signs of self-conscious linguistic innovation*" (Hodge 1992: 215). The phrase occurs then in the *Sketch* of 1842 (Darwin 1909: 7), "*but only once, late on, and with no accompanying sense of a special linguistic moment*", Hodge writes. By the mid 1850's, Darwin had decided to use "*natural selection*" as the title of his "*big book on species*" (Darwin 1975). It is only progressively that the metaphor of nature selecting is developed into an analogy, and "*a decisive element in the exposition of his theory*" (Hodge 1992: 215).

---

15   See *Notebook* D 135: "*The final cause of all this wedging, must be to sort out proper structure, & adapt it to changes*"; E 63: "*if nature had had the picking she would make them such a variety far more easily than man*". (Darwin 1987: 375–376, 414)

First, let us remember here that the title of the *Origin* is not the one originally chosen by Darwin who intended to name his book: "*An Abstract of an Essay on the Origin of Species and Varieties Through Natural Selection*" (Darwin to Lyell, 28 March 1859; Darwin 1991: 270). But this proposition was met with scepticism by Darwin's publisher John Murray who claimed, to Darwin's deep astonishment, that he did not understand the term. This is precisely why Darwin expands the title of the *Origin* with the phrase "*Through Natural Selection or the preservation of favoured races*", which he views as an "*explanation*" for natural selection (Darwin to Lyell, 30 March 1859; Darwin 1991: 273). If Darwin's own compatriots were so perplexed about the term's signification, it seems difficult to blame foreign translators for not understanding it either. Even Darwin himself laboured with the concept. He first described *natural selection* as "*a power incessantly ready for action*" and "*as immeasurably superior to man's feeble efforts, as the works of Nature are to those of Art*" (Darwin 1859: 61). Or in the terms of his letter to Asa Gray (5 September 1857):

> "I think it can be shown that there is such an unerring power at work, or Natural Selection (the title of my Book), which selects exclusively for the good of each organic being." (Darwin 1990: 447–448).

The exact scope of natural selection's power must constantly be defined: what does natural selection actually do? The following sentence is added to the third edition, close to the beginning of chapter 4: "*Several writers have misapprehended or objected to the term Natural selection. Some have even imagined that natural selection induces variability, whereas it implies only the preservation of such variations as occur and are beneficial to the being under its conditions of life*." (Darwin 1959: 164).

Strikingly enough, the problem of personification raised by Flourens, and the comparison with physics or chemistry evoked by Royer made their way into the *Origin*. As early as 1857 (29 November), Darwin wrote to Asa Gray:

> "I had not thought of your objection of my using the term 'natural selection' as an agent. I use it much as a geologist does the word Denudation—for an agent, expressing the result of several combined actions." (Darwin 1990: 492).

Darwin tried his best to avoid the personification. He invoked Newton's authority and asked: "*who objects to an author speaking of the attraction of gravity as ruling the movements of the planets?*" (Darwin 1959: 165). For Darwin, all scientific concepts are necessarily transfers or metaphors. The fate of Newtonian attraction in France is somewhat similar to that of Darwinian natural selection. After all, Newton had already been accused of the same fundamental sin: his "*attraction*" was supposed to be an occult quality under cover, pertaining to old Scholastic jargon and contrary to the French rationality of Cartesian physics. As Darwin states in the third edition:

> "It has been said that I speak of natural selection as an active power or Deity; but who objects to an author speaking of the attraction of gravity as ruling the movements of the planets? Every one knows what is meant and is implied by such metaphorical expressions; and they are almost necessary for brevity. So again it is difficult to avoid personifying the word Nature; but I

mean by Nature, only the aggregate action and product of many natural laws, and by laws the sequence of events as ascertained by us. With a little familiarity such superficial objections will be forgotten." (Darwin 1959: 165).

Darwin had difficulties accepting that his readers did not know how to interpret his new term. Was it the description of a force, a metaphor, or a personification of nature?

Wallace came back to this problem in a letter dated 2 July 1866—the fateful letter which finally convinced Darwin to adopt Herbert Spencer's expression "*survival of the fittest*" (Spencer 1864: 444, § 165)[16]. Wallace very vividly argued that "*natural selection*", although crystal clear for some readers, was nonetheless a stumbling block for many others. Here, Wallace was repeating to Darwin some complaints about the term gathered, among other sources, from an essay by the French philosopher Paul Janet (1823–1899)[17]. The fact that Wallace was clearly taking up Janet's critiques helps us understand that translations are not guilty of *spoiling* the original clarity of Darwin's thought. On the contrary, they reveal the shortcomings of the original. Accordingly, Wallace identified three different levels of difficulties with "*natural selection*".

A first group of concerns echoes Flourens' concern that the term natural selection suggests a personification of nature. Due to the parallel with man' selection, the term seems to require the constant watching of an intelligent *chooser* like man's selection to which it is so often compared by Darwin. But, more than that, it suggests that some sort of thought or direction is essential to the operation of natural selection, as if someone *had to* be thinking or directing the process. Therefore, the question remains open whether intention, that central motif of natural theology, has ever been excised out of the Darwinian framework. Robert Richards' (2009: 63–64) recent analysis of the Darwinian concept of natural selection goes as far as to suggest that this has never been the case and that Darwin could never get rid of intentions.

A second level of difficulty concerns the meaning(s) of natural selection within Darwin's text. What function does it really perform? Wallace singles out occurrences where natural selection seems to be synonymous with "*the simple preservation of favourable & rejection of unfavourable variations, in which case it is equivalent to* survival of the fittest"; and other instances where natural selection is taken "*for the effect or change, produced by this preservation*", as when Darwin refers to "*the circumstances favourable or unfavourable to natural selection*" and again when he thinks of isolation as "*an important element in the process of natural selection*" (Darwin 1859: resp. 84, 107, 104). Those last cases do not merely mean *survival of the fittest* but "*change produced by survival of the fittest*". A close reading

---

16   Spencer began issuing his *Principles of Biology* in fascicles in 1862.

17   The book of Janet (1864) re-issued two articles previously published in *La Revue des deux Mondes* (August and December 1863). There were several mediations between Darwin and Janet: Wallace was actually rephrasing a review of the English translation of the book (Janet 1866), published in *The Reader* (30 June 1866) (See Darwin 1985d: 230). Of course, when Wallace referred to "*natural selection*" in Janet's work, Janet had written "élection naturelle", relying on Royer's translation.

of the fourth chapter would document these recurring alterations of terms. Some even suggest that natural selection would be the direct cause for variations.

A third level of difficulty refers to the true meaning of natural selection in terms of its boundaries. In the 1860s, many readers of the *Origin* took *natural selection* as merely synonymous to the external conditions of life, the environmental cause or milieu placing demands on the varying individuals, choosing between them and hence bringing about evolutionary change. Thus understood as a cause or process rather than a mere outcome (and even if it strongly contradicts more recent claims on the Darwinian system), Darwin's natural selection may be a new way of naming Lamarck's milieu or environment. In this *"pruning"* conception of selection, variation is not understood as a necessary element of the process and selection is just the milieu working as a sieve.

Conscious of these difficulties, Darwin constantly attempted a response. His first operation was to distance himself from *natural selection*. As I mentioned earlier, Darwin successively suggested that natural selection was (1) used for the sake of brevity, (2) metaphorical, (3) a misnomer, and (4) a false term. Who knows why Darwin-the-man made those changes: maybe he was being too modest; maybe he was just fiddling with his text in order to dig out possible causes of misunderstandings; maybe he was misled in complying to his critics. But speculating on Darwin-the-man is just vain: we only have his texts to figure out what he meant. Hence, we have to pay attention to his rhetorical inflections, since they are all we have. In this case, they clearly indicate growing suspicion toward the therm.

Darwin's second lexical operation is to search for an equivalent that would offer the same meaning without the same linguistic traps: Spencer's *"survival of the fittest"* is one of them; *"preservation"* is another one, which he immediately used in the second half of the title of the book. Robert Young has aptly noted (1985: 95) *"that although the term 'preservation' eliminates some of the voluntarist overtones from the interpretation of the sources of variation, it still conveys the impression that active processes with voluntary overtones are operative in the accumulation of modifications."* Notwithstanding its own defects, the advantage of preservation is undoubtedly that it avoids any sense of an intelligent power acting in nature and Darwin refers to it in many letters in 1860 (Darwin to Bronn, 14 February; to W. Harvey, 20–24 September; to Gray, 26 September; to Lyell, 28 September—resp. Darwin 1993: 83, 371, 389, 397).

So why did Darwin stick to *"selection"*, in spite of all its faults? He gave at least four different sets of reasons. A first reason may sound purely emotional: he especially *"cared for it"*. But this is not sufficient reason and, as he wrote to Asa Gray, 11 May 1863 (Darwin 1999: 402–3): *"Personally, of course, I care much about natural selection, but that seems to me utterly unimportant compared to the question of* Creation *or* Modification*"*.

A second reason is grammatical: *"a great objection to* [the] *term"* [survival of the fittest], he wrote to Wallace, is *"that it cannot be used as a substantive governing a verb"* (Darwin to Wallace, 5 July 1866; Darwin 2004: 235). We have seen that similarly the problem of the French translators was not only to find an equivalent for *"selection"*, but to create a closely associated verb for *"to select"*.

A third reason is more purely theoretical. As he explained to Lyell (30 March 1859):

> "Why I like term is that it is constantly used in all works on Breeding, & I am surprised that it is not familiar to Murray; but I have so long studied such works, that I have ceased to be a competent judge." (Darwin 1991: 273).

Analogy seems to be the key to Darwin's stubbornness in using the *"false term"*. As he puts it to Wallace (5 July 1866), he had thought *"probably in an exaggerated degree, that it was a great advantage to bring into connection natural and artificial selection; this indeed led me to use a term in common, and I still think it some advantage"* (Darwin 2004: 235–6).

But a fourth reason can also be alleged: Darwin stuck to *"natural selection"* because the term was entrenched within his book and he couldn't tear it away without the whole structure of his *"Abstract"* falling apart. Undoubtedly natural selection had become a sort of structural constraint to the exposition of his thinking. He wrote to Wallace:

> "The term Natural selection has now been so largely used abroad and at home that I doubt whether it could be given up, and with all its faults I should be sorry to see the attempt made." (Darwin 2004: 236).

It is worth noting here that Darwin invoked translations as an incentive to stick with the original term. The argument is not rare in Darwin's prose. When V. Carus suggested that the German title of the book could be changed from Bronn's *Entstehung* into *Ursprung*, Darwin strongly rejected this proposition (21 June 1869):

> "I am very decidedly of opinion that it would be a mistake to change 'Entstehung' into 'Ursprung', although the latter may be much more correct. If this change were made, purchasers would think it a new book and would complain: I have known trouble thus caused in England." (Darwin 2009: 277).

Why resist a *"much more correct"* translation, if it was really so? Trapped in his own language, frightened that outraged buyers would rush to Down House and hang him for having sold the same book under two different titles, Darwin was condemned to optimism. Natural selection was part of his intellectual fate. He could only suggest to Wallace some final reasons to hope that *"in time, the term must grow intelligible"* and that in the end *"the objections to its use will grow weaker and weaker"* (Darwin 2004: 236).

However, Wallace himself would never be convinced, occasionally attacking publicly Darwin's metaphors as *"liable to misconception"* (Wallace 1870: 269) In his late recollections, re-telling the story of his *"Ternate episode"*, he clearly emphasized *"the efficient action of survival of the fittest in the improvement of the race"*, and he barely made use of the term *"natural selection"* (Wallace 1905: 363). As is clear from the *"Preface"* to his *Darwinism* (Wallace 1889), the agency of natural selection is certainly not subordinate to *"laws of variation, of use and disuse, of intelligence, and of heredity"*, as some of his contemporaries had it. But Wallace strongly emphasized that *"it has always been considered a weakness in Darwin's work that he based his theory, primarily, on the evidence of variation in domesticated animals and cultivated plants"*. By doing so, Wallace tacitly moved the theoretical centre of gravity from domestic to wild animals but also from natural selection to the study of variation.

CONCLUSION: HYDRA'S HEAD

Truly, if words are "*to the Anthropologist what rolled pebbles are to the Geologist–Battered relics of past ages often containing within them indelible records capable of intelligible interpretation*", as the astronomer John Herschel wrote in 1836 to the geologist Charles Lyell[18], this was certainly not the case with "*natural selection*." The term has been a constant puzzle for Darwin's readers. There are many ways of understanding Darwin's use of the phrase "*natural selection*" and Roger M. White (2010) has listed three of them:

(1) Darwin is speaking metaphorically when he talks of nature selecting.
(2) Darwin is introducing a new concept of selection that is applicable to both Nature and Man.
(3) The word "*select*" as applied to Nature is used in a special, technical, sense for use in biology, this sense being an analogical extension of the sense in human applications.

According to White (2010: 70–2), a close reading of the *Origin* "*shows Darwin oscillating between the first and the second of these three possibilities*" and the last one is only possible once a general theory of Natural selection is well established and understood.

It is quite obvious that, during the years following the publication of Darwin's concept of *natural selection*, Darwin's readers were lacking hindsight. Accordingly, it seems that much like Titus-Carmel's bananas, copies of the original multiplied notwithstanding that the original seemed doomed to inexorable decay. But after a few years, each illegitimate copy would wane and ultimately vanish until only carbon-copies of the original term were maintained. A similar pattern can be documented in many European languages, which cannot be accounted for solely by local circumstances: translators were first striving to find a local equivalent and ended up merely transliterating Darwin's term into their language. In French, Royer takes on Claparède's suggestion of using the phrase "*élection naturelle*"; but she ultimately has to abandon it in favour of *sélection naturelle*. Similarly, in Italian, Cannestrini's *elezione naturale* (Darwin 1864) was finally replaced by *selezione naturale* and, in Dutch, T.C. Winkler's *Natuurkeus* (literally: choice of nature, 1860*b-c*) was eventually abandoned and *natuurlijke selectie* is now wielded (for instance Darwin 2000). It seems that only later translations directly adopted a transliteration of *natural selection* like the first Spanish translation by Enrique Godínez (Darwin 1877), which refers to *la selección natural*.

To be sure, no Dogmatix miracle happened in the case of Darwin's "*natural selection*"; but the French and German translations of the original English phrase help us understand the various commitments of Darwin's concept, especially in

---

18   The letter is published by Cannon 1961.

relation to the breeders. As was bluntly remarked by C.L. Brace (1870: 287): "*The German term* 'Natural Breeding' *is not so good*". But contemporary English attempts were not more successful and Brace also rejected Spencer's "*survival of the fittest*". The reason he gave for his dissatisfaction with Spencer's phrase demonstrates the confusion about the concept of natural selection: according to Brace, survival of the fittest "*does not keep enough in view the ever-working forces proceeding under an intelligent plan, which we call Nature*". To be sure, Brace is strongly mistaken in his conception of Nature and in his reasons to praise and support "*natural selection*". But isn't it Darwin's tendency to personify the process that is largely responsible for this situation?

This conclusion will seem highly whimsical to biologists for whom terminological issues are just a pastime for philosophers. To be true, it now seems very obvious to biologists and others that there is no contradiction at all in such phrases as "*natural selection*" or "*unconscious choice*". If the analogy with artificial selection may have suggested consciousness to some readers, Darwin himself put more and more emphasis on "*unconscious selection*" in his account of the practice of breeders (Darwin 1859: 34–7). Nevertheless, it is important to keep those issues of agency and personification in mind, since they keep on recurring in Darwinian wars.

For instance, Peter Godfrey-Smith (2009) has recently argued that there are two approaches to natural selection. The first, classical (Darwinian) one, claims that "*evolution by natural selection results whenever there is a population in which there is variation between individuals, which leads to those individuals having different numbers of offspring, and which is heritable to some extent*" (Godfrey-Smith 2009: 4). Here, natural selection is a recipe, very much in the Darwinian manner: if such and such conditions are met, then natural selection necessarily follows. The second, more recent approach of natural selection deals with "*replicators*" (Dawkinsian). According to the latter, "*the first replicators arise at the origin of life itself, and are no more than single molecules. There is a raw evolutionary competition among these simple replicators; those that replicate faster and more accurately, and remain intact for longer, become more common.*" (Godfrey-Smith 2009: 5). For Godfrey-Smith, the main fault in the Dawkinsian approach is that it is "*agential*": it conceives of evolution "*in terms of a contest between entities with agendas, goals, and strategies*", the trouble being that, "*once we start thinking in terms of little agents with agendas—even in an avowedly metaphorical spirit—it can be hard to stop.*"

The historical and lexical analysis of this paper suggests that, if we take the term *natural selection* seriously, then it is agential and metaphorical through and through: agents are presupposed even in the recipe approach, even if it takes the guise of a mechanistic outcome. Quite obviously, the terms used to describe or name an entity, be they the Original or the translations, necessarily retro-act on the intelligibility of the entity, or the knowledge we take of it. And the issue with personification is not over in Darwinian studies[19].

---

19   As an example, see the recent polemics about Fodor & Piatelli-Palmarini's (2010) latest book, provocatively entitled *What Darwin Got Wrong*; and the replies of the Darwinian philosophers

It is not that all scientific concepts are necessarily metaphorical and Darwin at times claims that some concept will *"cease to be metaphorical"* (1859: 485). But, reading the *Origin*, one has to acknowledge that, if there is anything like *"Darwin's philosophy of science"*, it was not obsessed with clear definitions and purity of concepts. He easily accepted the metaphorical aspects of his concepts and used personification as a ploy to convey analogies. One can certainly conclude with Jon Hodge (1992: 212–3) that *"what really counted [for Darwin] was his argument for the analogy that the term was coined to signify, the analogy between man's selection and nature's"*. No matter what was Darwin's path to the discovery of natural selection, it is clear that he was happy to use the analogy, both for pedagogical and justificatory purposes: artificial selection brought to Darwin the kind of direct observational evidence that his philosophical masters (mostly Herschel, Whewell and Lyell) required to think about natural selection as a genuine Newtonian *vera causa*.

Accordingly, natural selection has produced a lot of *"illegitimate offspring"*, as Donna Haraway (1988) would put it, in the shape of directional concepts of evolution (Bowler 1988). After all, Darwin is not alone in this case and Lucretius in his time did not hesitate to personify nature: the word *foedera*, which is used in ordinary Latin to signify treatises between states, is attributed to nature in his *De Natura Rerum* (for instance, part I, verse 586; 2002: 122)*;* not to mention the idea of the *"laws of nature"* [*naturae leges*], another very successful and all-pervading personification of nature (e. g. Long 2005). Since Darwin's time, generations of evolutionary biologists have greatly refined Darwin's brilliant original insight, redefining it in relationship with mathematical *fitness*. Like many definitions, this one happens to create more difficulties than it solves. When biologists called in the tottering concept of fitness, hoping to finally get a (mathematical) grasp on natural selection, they basically traded one difficulty for another. Similarly, Montaigne (1989: 1046) had rejected the scholastic way of defining *homo* by *animal rationale* or *mortalis*, as making three difficulties out of one, with the following words: *"C'est la teste de Hydra."*

of science, for instance Michael Ruse (<http://chronicle.com/article/What-Darwins-Doubters-Get/64457/>); or Ned Block and Philip Kitcher who claim: *"Neither Darwin, nor any of his successors, believes in the literal* scrutiny *of variations. Natural selection, soberly presented, is about differential success in leaving descendants. If a variant trait (say, a long neck or reduced forelimbs) causes its bearer to have a greater number of offspring, and if the variant is heritable, then the proportion of organisms with the variant trait will increase in subsequent generations. To say that there is 'selection for' a trait is thus to make a causal claim: having the trait causes greater reproductive success."* (emphasis in the original) (<http://bostonreview.net/BR35.2/block_kitcher.php>).

ACKNOWLEDGEMENTS

This paper was first published as « Translating natural selection: true concept, but false term ? », in *Bionomina* 3 (2011) : 1–23. It is reprinted here with permission of Magnolia Press. I thank Alain Dubois (General Editor) and Zhi-Qiang Zhang (Managing Editor) for permission. Originally, the idea of this paper budded off my *Habilitationsschrift* (Paris, 2007) published in 2009 as *"Darwin contre Darwin"*. It is part of the BIOSEX project, funded by the French ANR (Agence nationale de la Recherche, ANR-07-JCJC-0073–01). A first presentation was made at Virginia Tech (Nov. 4, 2009) thanks to the generous support of Richard Burian. I completed the paper during a short-term stay as a Visiting Research Fellow at the Philosophy Department of the University of Leeds (February–March 2010): lots of improvements are due to comments, critiques and suggestions of the energetic Leeds HPS Team, especially to Greg Radick and Jon Hodge. The final draft was greatly improved thanks to the insightful comments and kind observations of two anonymous referees, whom I thank here. I thank Claude-Olivier Doron for drawing relevant passages of Gayot and Lefebvre Sainte-Marie to my attention.

REFERENCES

Anonymous (1860) 'Sélection naturelle. Choix de la nature', Magasin pittoresque 28 (09): 294–6.

Beddall, B.G. (1988) 'Wallace's annotated copy of Darwin's Origin of species', Journal of the History of Biology 21/22: 265–89.

Beer, G. (1996) 'Introduction. Note on the text' in C. Darwin, On the origin of species (Oxford: University Press): vii–xxix.

Bernard, C. (1865/1984) Introduction à l'étude de la médecine expérimentale (Paris: Flammarion): 1–320.

Bock, W.J. (2009) 'The Darwin-Wallace Myth of 1858' Proceedings of the zoological society 62 (1): 1–12.

Bowler, P.J. (1983) The eclipse of Darwinism (London & Baltimore: Johns Hopkins University Press).

Bowler, P.J. (1988) The non-Darwinian revolution. Reinterpreting a historical myth. (London & Baltimore: Johns Hopkins University Press).

Brace, C.L. (1870) 'Darwinism in Germany', North American Review 110–112: 284–299.

Brandon, R. & R. Burian (ed) (1984) Genes, organisms and populations. Controversies over the units of selection (Cambridge, MA: Bradford & MIT).

Bronn, H.G. (1860a) 'Ch. Darwin, On the origin of species', Neues Jahrbuch für Mineralogie 35: 112–6.

Bronn, H.G. (1860b) 'Schlusswort des Uebersetzers', in Darwin, C. (1860) Ueber die Entstehung der Arten…, (Stuttgart: Schweitzerbart): 495–520.

Browne, J. (2002) Charles Darwin. Vol. 2. The power of place. The Origin and after. The years of fame (Princeton, NJ: University Press).

Büchner, L. (1869) 'Sechs Vorlesungen über die Darwin'sche Theorie, T. Thomas, Leipzig', French translation by A. Jacquot, Conférences sur la théorie darwinienne de la transmutation des espèces et de l'apparition du monde organique (Paris: Reinwald).

Cannon, W.F. (1961) 'The Impact of Uniformitarianism: Two Letters from John Herschel to Charles Lyell, 1836–1837', Proceedings of the American Philosophical Society 105 (3): 301–14.

Claparède, E. (1861) 'M. Darwin et sa théorie de la formation des espèces', Revue germanique 16: 523–59.

Conry, Y. (1974) L'introduction du darwinisme en France au XIXe siècle (Paris: Vrin).

Darwin, C. (1859) On the origin of species by means of natural selection, or the preservation of favoured races in the struggle for life (London: Murray).

Darwin, C. (1860a) Ueber die Entstehung der Arten im Thier- und Pflanzenreich durch natürliche Zuchtung, oder Erhaltung der vervollkommneten Rassen im Kampfe um's Dasein (Stuttgart: Schweitzerbart).

Darwin, C. (1860b) *Het ontstaan der soorten van dieren en planten door middel van de natuurkeus, of het bewaard blijven van bevoorregte rassen in den strijd des levens*, Volume 1 (Haarlem: A.C. Kruseman).

Darwin, C. (1860c) *Het ontstaan der soorten van dieren en planten door middel van de natuurkeus, of het bewaard blijven van bevoorregte rassen in den strijd des levens*, Volume 2 (Haarlem: A.C. Kruseman).

Darwin, C. (1862) De l'origine des espèces, ou Des lois du progrès chez les êtres organisés, traduit en français sur la troisième édition (Paris: Guillaumin-Masson).

Darwin, C. (1864) Sulla origine delle specie per elezione naturale, owero Conservazione delle razze perfezionate nella lotta per l'esistenza, prima traduzione italiana col consenso dell'autore, per cura di Giovanni Canestrini e Luigi Salimbeni (Modena: Tipi di Nicola Zanichelli e Soci).

Darwin, C. (1866) De l'origine des espèces par sélection naturelle, ou Des lois de transformation des êtres organisés. Deuxième édition française, revue sur la quatrième édition anglaise (Paris: Guillaumin & Masson).

Darwin, C. (1868) The variation of animals and plants under domestication, Vol. 2 (London: Murray).

Darwin, C. (1873) L'origine des espèces au moyen de la sélection naturelle, ou La lutte pour l'existence dans la nature. Traduit sur l'invitation et avec l'autorisation de l'auteur sur les cinquième et sixième éditions anglaises. Augmentées d'un nouveau chapitre et de nombreuses notes et additions de l'auteur, par J.-J. Moulinié (Paris: C. Reinwald & Cie).

Darwin, C. (1876a) L'origine des espèces au moyen de la sélection naturelle, ou La lutte pour l'existence dans la nature. Traduit sur la sixième édition anglaise par E. Barbier (Paris: C. Reinwald & Cie).

Darwin, C. (1876b) Über die Entstehung der Arten durch natürliche Zuchtwahl oder die Erhaltung der begünstigten Rassen im Kampfe um's Dasein. Aus dem Englischen übersetzt von H.G. Bronn. Nach der sechsten englischen Auflage wiederholt durchgesehen und berichtigt von J. Victor Carus. 6. Auflage (Stuttgart: Schweizerbart).

Darwin, C. (1877) *Orígen de las especies por medio de la selección natural ó La conservación de las razas favorecidas en la lucha por la existencia. Traducida con autorizacion del autor de la sexta y última edicion inglesa, por Enrique Godinez* (Madrid-Paris: Biblioteca Perojo).

Darwin, C. (1959) The origin of species. A variorum text, edited by Morse Peckham (Philadelphia: University of Pennsylvania Press).

Darwin, C. (1975) Charles Darwin's Natural selection, being the second part of his big species book written from 1856 to 1858. Edited by R.C. Stauffer (Cambridge, MA: University Press).

Darwin, C. (1990) The correspondence of Charles Darwin. Edited by F. Burkhardt & S. Smith. Volume 6 (Cambridge, MA: University Press).

Darwin, C. (1991) The correspondence of Charles Darwin. Edited by F. Burkhardt & S. Smith. Volume 7 (Cambridge, MA: University Press).

Darwin, C. (1993) The correspondence of Charles Darwin. Edited by F. Burkhardt, D.M. Porter, J. Browne & M. Richmond. Volume 8 (Cambridge, MA: University Press).

Darwin, C. (1999) The correspondence of Charles Darwin. Edited by F. Burkhardt, D.M. Porter, S.A. Dean, J.R. Topham & S. Wilmot. Volume 11 (Cambridge, MA: University Press).

Darwin, C. (2004) The correspondence of Charles Darwin. Edited by F. Burkhardt, D.M. Porter, S.A. Dean, S. Evans, S. Innes, A. Sclater, A. Pearn & P. White Volume 14 (Cambridge, MA: University Press).

Darwin, C. (2009) The correspondence of Charles Darwin. Edited by F. Burkhardt, J.A. Secord, S.A. Dean, S. Evans, S. Innes, A.M. Pearn & P. White. Volume 17 (Cambridge, MA: University Press).

Darwin, C. (1987) Charles Darwin's notebooks, 1836–1844: geology, transmutation of species, metaphysical enquiries. Transcribed and edited by P.H. Barrett, P.J. Gautrey, S. Herbert, D. Kohn, S. Smith (Ithaca, NY: British Museum & Cornell University Press).

Darwin, C. (2000) Over het ontstaan van soorten door middel van natuurlijke selectie, of het behoud van bevoordeelde rassen in de strijd om het leven. Dutch translation by Ludo Hellemans (Amsterdam: Nieuwezijds).

Darwin, F. (ed) (1909) *The foundations of the origin of species. Two essays written in 1842 and 1844* (Cambridge, MA: Cambridge University Press).

Derrida, J. (1967) De la grammatologie (Paris: Minuit).

Derrida, J. (1978) La vérité en peinture (Paris: Flammarion).

Ellegård, A. (1958/1990) Darwin and the general reader: the reception of Darwin's theory of evolution in the British periodical press, 1859–1872 (Chicago, IL: University Press).

Elshakry, M.S. (2008) 'Knowledge in motion. The cultural politics of modern science translations in Arabic', Isis 99: 701–730.

Flourens, P. (1864) Examen du livre de M. Darwin sur l'origine des espèces (Paris: Garnier).

Fodor, J. & M. Piatelli-Palmarini (2010) What Darwin got wrong (New York: Farrar, Straus & Giroux).

Gayon, J. (1998) Darwinism's struggle for survival. Heredity and the hypothesis of natural selection. Translation Matthew Cobb (Cambridge, MA: University Press).

Gayon, J. (2009) 'Darwin et Wallace: un débat constitutif pour la théorie de l'évolution par sélection naturelle', in L'évolution aujourd'hui à la croisée de la biologie et des sciences humaines, Actes du colloque des 29, 30 et 31 janvier 2009 à l'Académie royale de Belgique (Bruxelles: Académie royale de Belgique): 89–122.

Gayon, J. & D. Zallen (1998) 'The role of the Vilmorin Company in the promotion and diffusion of the experimental science of heredity in France', 1840–1920, Journal of the History of Biology 31(2): 241–62.

Gayot, E. (1850) La France Chevaline (Paris: Comptoir des imprimeurs unis et Vve Bouchard-Huzard).

Gliboff, S. (2008) H.G. Bronn, Ernst Haeckel, and the origins of German Darwinism. A study in translation and transformation (Cambridge, MA: MIT).

Glick, T.F. (1974) The comparative reception of Darwinism (Austin: University of Texas Press).

Godfrey-Smith, P. (2009) Darwinian populations and natural selection (Oxford: Oxford University Press).

Haraway, D.J. (1988) 'Situated knowledges: The science question in feminism and the privilege of partial perspective', Feminist Studies 14(3): 575–99.

Hergé [Remi , G.P.] (1964/1993) The broken Ear. Translation L, Lonsdale-Cooper & M. Turner (Tournai: Casterman).

Hodge, M.J.S. (1992) 'Natural selection, historical perspectives', in E.F. Keller & E.A. Lloyd (ed), Keywords in evolutionary biology (Cambridge, MA: University Press): 213–9.

Hodge, M.J.S. & D. Kohn (1985) 'The immediate origins of natural selection', in D.Kohn (ed), The Darwinian heritage (Princeton, NJ: University Press): 185–206.

Hoquet, T. (2009) Darwin contre Darwin. Comment lire l'Origine des espèces? (Paris: Le Seuil).

Huneman, P. (2009) 'Sélection', in: T. Heams et al. (ed), Les mondes darwiniens: l'évolution de l'évolution (Paris: Syllepse): 47–86.

Janet, P. (1864) Le matérialisme contemporain en Allemagne. Examen du système du docteur Büchner (Paris: Baillière).

Janet, P. (1866) The materialism of the present day. A critique of Dr Büchner's system. Translated fron the French by Gustave Masson (London: H. Baillière).

Johnson, C.N. (2007) 'The Preface to Darwin's Origin of species: the curious history of the historical Sketch', Journal of the History of Biology 40: 529–56.

Kant, I. (2007) Critique of pure reason. English translation N. Kemp Smith. Revised second edition. (Houndmills-Basingstoke, Hampshire: Palgrave Macmillan).

Keller, L. (ed) (1999) Levels of selection in evolution (Princeton, NJ: University Press).

Kohn, D. (ed) (1985) The Darwinian heritage (Princeton, NJ: University Press).

Lefebvre-Sainte-Marie, G. (1849) De la Race bovine courte corne améliorée, dite race de Durham en Angleterre, aux États-Unis d'Amérique et en France (Paris: Impr. Nationale).

Lehoux, D. (2006) 'Laws of nature and natural laws', Studies in History and Philosophy of science 37: 527–49.

Limoges, C. (1970) La sélection naturelle, étude sur la première constitution d'un concept (1837–1859) (Paris: Presses Universitaires de France).

Long, A.A. (2005) 'Law and nature in Greek thought', in M. Gagarin et al. (ed), The Cambridge companion to ancient Greek law (Cambridge, MA: University Press): 412–30.

Lucas, P. (1847a) Traité philosophique et physiologique de l'hérédité naturelle dans les états de santé et de maladie du système nerveux, avec l'application méthodique des lois de la procréation au traitement général des affections dont elle est le principe, Volume 1 (Paris: Baillière).

Lucas, P. (1847b) Traité philosophique et physiologique de l'hérédité naturelle dans les états de santé et de maladie du système nerveux, avec l'application méthodique des lois de la procréation au traitement général des affections dont elle est le principe, Volume 2 (Paris: Baillière).

Lucretius (2002) De la nature des choses, trad. par Bernard Pautrat (Paris: Librairie générale française).

Matthew, P. (1831) On naval timber and arboriculture; with critical notes on authors who have recently treated the subject of planting (London: Longman, Rees, Orme, Brown, and Green).

Matthew, P. (1860) 'Nature's Law of Selection', The Gardener's Chronicle, 7 April: 312–3.

Miles, S.J. (1889) 'Clémence Royer et De l'origine des espèces, traductrice ou traîtresse?', Revue de Synthèse (4)1: 61–83.

Montaigne, M. de (1989) Œuvres complètes, textes établis par A. Thibaudet et M. Rat (Paris: Gallimard, Bibliothèque de la Pléiade).

Montgomery, S.L. (2000) Science in translation, movements of knowledge through cultures and time (Chicago, IL: University Press).

Naudin, C.V. (1852) 'Considérations philosophiques sur l'espèce et la variété', Revue Horticole, 4e série(1): 102–9.

Okasha, S. (2006) Evolution and the levels of selection (Oxford: Clarendon Press).

Owen, R. (1860) 'Review of Origin & other works', Edinburgh Review 111: 487–532.

Plate, L.H. (1903) Über die Bedeutung des Darwin'schen Selectionsprincips und Probleme der Artbildung (Leipzig: W. Engelmann).

Richards, R.J. (2009) Darwin's theory of natural selection and its moral purpose, in R.J. Richards & M. Ruse (ed), The Cambridge companion to The origin of species (Cambridge, MA: University Press): 47–66.

Rosset, C. (1977) Le réel: traité de l'idiotie (Paris: Minuit).

Royer, C. (1862) 'Préface du traducteur', in Darwin (1862): v-lxiv.

Royer, C. (1866) 'Avant-propos', in Darwin (1866): i-xiii.

Rupke, N. (2000) 'Translation studies in the history of science. The example of Vestiges', The British Journal for the History of Science 33: 209–22.

Ruse, M. (1975) 'Darwin and artificial selection', Journal of the History of Ideas 36 (2): 339–50.

Sebright, J.S. (1809) The art of improving the breeds of domestic animals. Reprinted in pages 93–126 of C.J. Bajema, Artificial selection and the development of evolutionary theory (Strougsburg, PA: Huchinson Ross).

Seidlitz, G. (1871) Die Darwins'che Theorie, elf Vorlesungen über die Entstehung der Thiere und Pflanzen durch Naturzüchtung (Dorpat: E. Mattiesen).

Sinclair, J. (1821) The code of agriculture; including observations on gardens, orchards, woods and plantations. Third edition (London: Sherwood).

Sinclair, J. (1824) L'agriculture pratique et raisonnée. Traduit de l'anglais par C.-J.-A. Mathieu de Dombasle. 2 volumes (Paris: Mme Huzard-Déterville).

Sober, E. (1984) The nature of selection. Evolutionary theory in philosophical focus (Chicago, IL: University Press).

Sober, E. & D.S. Wilson (1998) Unto others: the evolution and psychology of unselfish behavior. (Cambridge, MA: University Press).

Spencer, H. (1864) The Principles of biology, Volume 1 (New York: Appleton).

Todes, D.P. (1989) Darwin without Malthus. The struggle for existence in Russian evolutionary thought (New York: Oxford University Press).

Venuti, L. (1995) The translator's invisibility: a history of translation (London & New York: Routledge).

Vilmorin, L. (1859) Notices sur l'amélioration des plantes par le semis et considérations sur l'hérédité dans les végétaux (Paris: Librairie Agricole).

Wallace, A.R. (1858) 'On the tendency of varieties to depart indefinitely from the Original Type', in C.R. Darwin & A.R. Wallace, On the tendency of species to form varieties; and on the perpetuation of varieties and species by natural means of selection. [Read 1 July], Journal of the Proceedings of the Linnean Society of London, Zoology 3: 45–62.

Wallace, A.R. (1870) Contributions to the theory of natural selection: a series of essays (London: Macmillan).

Wallace, A.R. (1889) Darwinism. An exposition of the theory of natural selection with some of its applications (London: Macmillan).

Wallace, A.R. (1905) My life. A record of events and opinions. Vol. 1. (London: Chapman et Hall).

Weismann, A.F.L. (1886) Uber die Bedeutung der sexuelle Fortpflanzung für die Selektionstheorie (Jena: G. Fischer).

Weismann, A.F.L. (1893) Die Allmacht der Naturzüchtung. Eine Erwiderung an Herbert Spencer (Jena: G. Fischer).

Weismann, A.F.L. (1909) Die Selektionstheorie, eine Untersuchung (Jena: G. Fischer).

White, R.M. (2010) Talking about God, the concept of analogy and the problem of religious language (Farnham, Surrey, England: Ashgate).

Wilberforce, S. (1860) 'Darwin's Origin of species', Quarterly Review 108: 225–64.

Youatt, W. (1834) A history of the horse, in all its varieties and uses, together with complete directions for the breeding, rearing, and management (Washington: Duff Green).

Youatt, W. (1840) Sheep, their breeds, management and diseases (London: Baldwin and Cradock).

Youatt, W. (1851) Le cheval, traduit de l'ouvrage anglais The horse (…), par H. Cluseret (Paris: Dentu).

Young, R.M. (1985) Darwin's metaphor. Nature's place in Victorian culture (Cambridge, MA: Cambridge University Press).

Zirkle, C. (1941) 'Natural selection before the Origin of species', Proceedings of the American philosophical Society 84: 71–123.

Zuber, Z. (1968) Les "Belles Infidèles" et la formation du goût classique. Perrot d'Ablancourt et Guez de Balzac (Paris: Armand Colin).

THE MAPPING OF HUMAN BIOLOGICAL AND LINGUISTIC DIVERSITY: A BRIDGE BETWEEN THE SCIENCES AND HUMANITIES[1]

Frank Kressing

INTRODUCTION

The late 20th century saw a proliferation of presumably new approaches claiming a direct link between the evolution of linguistic diversity and biodiversity in humans.[2] In this paper, I show that the so-called 'new synthesis' of genetic, linguistic, and archaeological data popularised since the 1980s (e.g. Cavalli-Sforza et al. 1988, 1994, 2001, 2003) is based on well-established and deeply rooted traditions origination in European scholarship, as thorough reviews of the available literature have indicated (e.g. Alter 1999; Archibald 2009). Based on an analysis of the historical development of evolutionary and co-evolutionary theories, I emphasise the existence of an interdisciplinary network encompassing the fields of biology and linguistics in the 18th–20th centuries (see Krischel, this volume) and focus on the concepts and ideas transferred through these interpersonal reticulations, arguing that interpersonal networks transgressing the boundaries between science and the humanities contributed significantly to the formation of evolutionism as the leitmotif of the 19th century (Krischel, Kressing and Fangerau 2011; Kressing 2012; Kressing, Fangerau and Krischel 2013).

Numerous authors have examined the transmission of evolutionary theories between scholars in linguistics and biology (Bock 1955, Sahlins 1976; Hull 1988; Bowler 1983; Boyd and Richardson 1985; Römer 1989; Marks 1995; Trigger 1998; Alter 1999; Sahlins 2000; Atkinson and Gray 2005; Hutton 2005; Archibald 2008; Desmond and Moore 2009; Ragan 2009; Dux 2011). These authors have characterised the situation as follows:

(1) The idea of evolution developed within close interdisciplinary networks encompassing the sciences and humanities.

---

1   A number of ideas outlined in this contribution have also been presented in Kressing, Fangerau & Krischel (2013) which is a paper that is primarily concerned with past and present attempts in the 'biologisation' of cultural and linguistic differences. Contrary to that publication, this chapter focuses on interdisciplinary networks in the sciences and the humanities in the formulation of a unified theory of evolution and thus presents a topic different from Kressing, Fangerau & Krischel (2013).

2   For recent accounts, see Gray and Atkinson (2003), Gray (2005), Greenhill, Blust and Gray (2008), and – an attempt to mathematically model cultural evolution – Currie, Greenhill, Gray, Hasegawa and Mace (2010).

(2) Linguistics had a decisive influence on the formulation of an evolutionary theory applied to humans (cf. Römer 1989; Atkinson and Gray 2005; Hutton
2005).

(3) The idea of staged hierarchical development developed first in the social sciences, then in linguistics, and, finally, in biology. The idea of evolution is rooted
in pre-Darwinian, Enlightenment-period thinking and reflects European ideas
of superiority and inferiority defined by stages of progress in a hierarchical
scheme (Sahlins 1976: 93–107; Hutton 2005: 21). As these pre-Darwinian con-
ceptions of evolution were bound to teleology, development was identified with
improvement.

(4) Scholars in biology and linguistics favoured models of unilinear descent according to pedigree, which emphasised 'pure blood lines' instead of lateral or
horizontal transfer among languages, species, and 'races'. Purity of descent
was venerated as the path to evolutionary perfection, whereas the mixture of
languages, species, and 'races' was perceived as degeneration.

In contrast to recent scholarship's emphasis on pure unilinear descent without
admixture, I would like to highlight a second narrative that has emerged from a
historiographical perspective on evolution, namely the formulation of the idea of
evolution through the lateral transfer of ideas and reticulations between scholars in
the sciences and humanities on intellectual and personal levels. Going beyond these
currently circulating competing theses, I argue that interdisciplinary discourse centred on evolution evolved from the 16th-century foundation of anthropology as 'the
study of man's double nature', which has had a lasting influence that persists to the
present day.[3]

In support of my argument, I first consider the prevailing traditions in the classification of human languages and human biodiversity, then turn to concepts of
evolution and co-evolution, and finally argue that theories of a unified human evolution are rooted in the historical development of the discipline of anthropology,
stressing the legacy of anthropology as the intellectual background for interdisciplinary scientific transfer. Thus, thinking in terms of co-evolution can be perceived as
a legacy of the establishment of anthropology as an academic discipline.

3   Anthropology as the 'science of man' originated during Renaissance times as an attempt to li-
    berate science from theology. At the turn from the sixteenth to the seventeenth centuries, Otto
    Cassmann (1562–1607) described 'anthropology' as 'the lore of human nature' which – accor-
    ding to him – constituted a 'double form of existence, being bound to the world's spiritual as
    well as to the world's physical essence' (Cassmann 1594). This can be seen as the main reason
    why the term 'anthropology' for a long time maintained a two-folded meaning, relating to the
    physical as well as to the mental sphere of human existence.

COMMON TRADITIONS IN THE MAPPING OF HUMAN LINGUISTIC AND BIOLOGICAL DIVERSITY

The first attempts to map human linguistic diversity can be traced to the 16th century. The identification of different linguistic stocks, such as Uralic (Witsen 1692; Gyarmathi 1799) and Indo-European (Boxhorn 1647 a, b, c; Jones 1786), resulted in the development of phylogenetic models of language descent, with the first 'trees of language' emerging around 1800 (e.g. Gallet 1795/1800; Schleicher 1853, 1861/62; cf. Auroux 1990; McMahon 2004, McMahon & McMahon 2010; Driem 2005).

The first phylogenetic models of biological descent developed at approximately the same time (Augier 1801; cf. Stevens 1983), although the 'language tree' metaphor was accepted earlier than the 'tree of life' concept. Human phenotypes were classified using schemes containing three to twelve different 'races' (Haeckel 1874; cf. Brues 1977; Schwidetzky 1992; Marks 1995) in a taxonomic effort that was part of the research agenda of physical anthropology, an academic field that emerged in the 18th century (Hoßfeld 2005).[4]

These 18th- and 19th-century 'racial' and linguistic classifications not only provided linguistic and biological taxonomies, but also established hierarchical schemes intending to show the development from 'lower' to 'higher races' (Meiners 1785; Gobineau 1853/55; cf. Römer 1989; Hutton 2005) and from 'inferior to 'superior' languages (Schlegel 1808; Humboldt 1820, 1836; cf. Sapir 1921). In other words, they depicted progressive human development and increasing complexity (Spencer 1857) in accordance with evolutionism as the leitmotif of the 19th century.

The idea of the co-evolution of biological features and languages can be traced to the beginning of the Enlightenment (Bock 1955: 133).[5] In the 18th century, Adam Ferguson (1767) in Scotland and Antoine de Condorcet (1795) in France described progressive social development through three stages, from savagery through barbarism to civilization (Trigger 1998: 32). In the 19th century, this three-stage model inspired Henri de Saint Simon (1760–1825) in France, Gustav Klemm (1802–1867) in Saxony, and Lewis Henry Morgan (1818–1881) in the United States to promote models of universal cultural evolution.[6] Morgan was also strongly influenced by the

---

4   Main representatives of early physical anthropology were Georges-Louis de Buffon (1707–1788), Friedrich Blumenbach (1752–1840), Lorenz Oken (1779–1851), Immanuel Kant (1724–1804), Gottfried Herder (1744–1803), and Georg Forster (1754–1794). It should be pointed out that these founding fathers of physical anthropology did not restrict their education to the medical field: Buffon was a mathematician and botanist as well as a physician; Forster is known as an ethnologist and journalist as well as a naturalist; Kant included philosophy, physics, mathematics, and sciences in his studies; and Oken, trained as a physician, later shifted his attention to zoology and natural philosophy. These wide-ranging fields of study clearly demonstrate that academic disciplinary borders were much less solid than today and were often transgressed by these early representatives of anthropology.

5   When Jean-Baptiste Lamarck (1809), Robert Chambers (anonymously, 1844), and Charles Darwin (1859) published their contributions to biological evolution, they relied on an established tradition of evolutionary thinking in natural history and sociology.

6   All of these authors distinguished – in different manners – three stages of evolution: slavery,

Swiss lawyer Jakob Bachofen's (1815–1887) theory of matriarchy (1861; cf. Rössler 2007: 5) and he, in turn, had a decisive influence on the formation of Friedrich Engels' (1820–1895) and Karl Marx's (1818–1883) theory of revolutionary social change.[7]

Another lineage of intellectual inspiration runs from Condorcet and the Belgian statistician Adolphe Quételet (1796–1874) to the French philosopher August Comte (1798–1857), co-founder and denominator of the discipline of sociology. Comte introduced a historical perspective to the study of human societies,[8] and Darwin (1809–1882) historicised nature approximately two decades later in *On the Origin of Species* (1859). In this text, Darwin adopted the term 'survival of the fittest' from the sociologist Herbert Spencer (1820–1903), who had employed Comte's term '*sociologie*' (Spencer 1857) and later used Darwin's term 'natural selection' (Alter 1999: 23, 30; Beer 1971: 565–577).

The prevalence of close mutual relations among the proponents of evolutionism, especially in the mid-19th century, is further illustrated by the relationships among the biologist Ernst Haeckel (1834–1919), the foremost populariser of Darwinism in German-speaking areas, the linguist August Schleicher (1821–1868), and his colleague Wilhelm Bleek (1827–1875). Bleek, who was Ernst Haeckel's son in law (Koerner 1983: xi), and Schleicher favoured a pedigree model of language origins that fitted perfectly with the 'tree of life' tentatively envisaged by Darwin and elaborately visualised by Haeckel (Schleicher 1863). Haeckel applied the theory of biological evolution to humans in a far more pronounced way than Darwin. Schleicher directly equated languages with species and dialects with subspecies (cf. Schleicher 1863; Suttrop 1999; Koerner 1983; Atkinson and Gray 2005; Uschmann 1972).

Among Anglophone scholars, ideas of linguistic evolution were transmitted to Charles Lyell (1797–1875), Asa Gray (1810–1888), and Thomas Huxley (1825–1895), members of Darwin's 'inner circle', by Darwin's cousin and brother in law, Hensleigh Wedgewood (1803–1891; Desmond and Moore 2009: 57),[9] and by the German-born linguist Max Müller's (1823–1990) significant contributions to the post-*Origin* debate. Both of these influences may explain the frequent references to linguistics in Darwin's work.[10]

---

feudalism, and civic society (Saint-Simon); savagery, domestication, and freedom (Klemm); and savagery, barbarism, and civilization (Morgan).

7   Friedrich Engels' *The Origin of the Family, Private Property, and the State* (1884) is subtitled '*Subsequent to Lewis Henry Morgan's Research*' (in German: *Der Ursprung der Familie, des Privateigenthums und des Staats. Im Anschluss an Lewis H. Morgans Forschungen*).

8   He perceived societies as entities that undergo 'ageing', meaning development and change.

9   Thomas Huxley was fluent in German; he reviewed und published Schleicher's *Die Darwinsche Theorie und die Sprachwissenschaft* (1863) in Britain (Alter 1999: 79).

10 E. g. *On the Origin* ... (Darwin 1859: 342). Several authors (Alter 1999; Marks 1995; Römer 1989; Hutton 2005) have indicated that the idea of an intrinsically connected co-evolution of languages and so-called 'races' had become an established paradigm in the second half of the 19th century, when originally purely linguistic designations such as Aryan or Semitic were increasingly associated with 'racial' affiliations (Müller 1855, 1861; Lapouge 1899; LeBon 1894).

THE ORIGIN AND DEVELOPMENT OF ANTHROPOLOGY

After having pointed out that the theory of social, linguistic, and biological human co-evolution was shaped through close intellectual and personal reticulations among influential scholars in different disciplines, I argue that one reason for the establishment of these networks can be found in the history of the scientific discipline of anthropology, and particularly in its genuinely interdisciplinary aspects (Dux 2011: 42; Streck 2000: 141).[11]

Fostered by the 17th-century Cartesian dualism of body and soul,[12] anthropology split into the fields of physical and cultural anthropology. Cultural anthropology was established in the 18th century under designations such as *Völkerkunde* (Schlözer 1772), *ethnographie* (Gatterer 1775), and *ethnologie*.[13] The first ethnographic societies, founded in the 19th century,[14] were not clearly distinguished from anthropological societies and had interdisciplinary memberships recruited from the sciences and humanities. One example is the *Berliner Gesellschaft für Völkerkunde, Ur- und Frühgeschichte* (Berlin Society for Ethnography, Prehistory and Early History) founded in 1869 by the physicians Rudolf Virchow (1821–1902) and Adolf Bastian (1826–1905).

Due to efforts by Bastian in Germany and Edward Tylor (1832–1917) in Britain, cultural anthropology was established as an independent academic discipline in European universities.[15] In North America, however, Bastian's disciple Franz Boas (1858–1942)[16] perpetuated the tradition of integrating physical and cultural anthropology by establishing the so called 'four field approach',[17] which also included the anthropological sub-disciplines of archaeology and linguistics.[18]

11   Many authors have thoroughly investigated the origins of anthropology (e.g. Leclérc 1972; Stagl 1974; Girtler 1979; Stocking 1988; Streck 2000; Barth 2005; Gingrich 2005; Hann 2005; Dux 2011).
12   Although Platonic philosophy considered the dichotomy between body and soul, Cartesian dualism renewed interest in this problem in the 17th century and advanced the split of anthropology into two distinct branches (Descartes 1641; Zittel 2009).
13   The term '*Ethnologie*' was first used in 1783 in Vienna (Rössler 2007) and in 1787 by the Swiss philosopher Alexandre César Chavannes (1731–1800; Bitterli 1991).
14   The *Société des Observateurs de l'Homme* (1799), of which Jean Baptiste Lamarck was a member, *Société d'Ethnologie* (1839), Ethnological Society of London (1844), and Anthropological Society (1839). The latter two were merged to form the Royal Anthropological Society in 1871.
15   In Berlin in 1869 and Oxford in 1884 (by Edward Tylor; cf. Rössler 2007: 6). In the beginning of the 20th century, four main traditions of cultural anthropology emerged: social anthropology in Britain (cf. Malinowski 1915, 1922; Radcliff-Brown 1922; Evans-Pritchard 1937), *ethnologie* in France, *Völkerkunde* in German-speaking countries, and cultural anthropology in North America (Hann 2005).
16   Franz Boas emigrated from Germany to the United States in 1886 and was appointed Chair of Anthropology in Worcester, Massachusetts, in 1888.
17   The 'four field approach', which refers to the inclusion of physical anthropology, cultural anthropology, linguistics, and archaeology within the framework of 'anthropology' as an integrative scholarly discipline, was established in the academic tradition of the United States and is tied to Franz Boas, considered to be the founding father of North American anthropology.
18   Expressed by the founding of the American Anthropological Association (1902) and the journal

THE DECLINE OF EVOLUTIONISM IN THE BEGINNING
OF THE 20TH CENTURY

Boas is known foremost for his strong anti-evolutionary, particularistic, and relativistic approach, which shaped a generation of anthropologists in North America and accomplished the 'divorce of race and culture' (Marks 1995: 71). He claimed that culture, 'race', and language constituted mutually independent and unrelated determinants of human existence (Boas 1913, 1940; cf. Streck 2000: 142). This anti-evolutionist orientation of the Boasian school coincides with the general dismissal of evolutionist thinking in biology, anthropology, sociology, and linguistics in the beginning of the 20th century. All four traditions (Hann 2005) of cultural anthropology in Anglo-Saxon, French, and German-speaking countries expressed a strong anti-evolutionary perspective, guided by the theories of cultural relativism and particularism, diffusionism, (Frobenius 1898; Graebner 1905: 28–53, 1911; Schmidt 1912–55; Koppers 1915–16; cf. Streck 2000: 43)[19] and structuralism.[20] Bowler (1983) provided evidence for the fostering of this simultaneous decline of evolutionism by, among other factors, interpersonal networks, such as those among Boas, Ferdinand de Saussure (1857–1913), Claude Lévi-Strauss (1908–2009), and Roman Jakobson (1896–1982; Parkin 2005: 209–210).[21] Thus, ideas of diffusionism

American Anthropologist (already in 1888, the 'new series' under the editorship of F. Boas started in 1899). This interdisciplinary framework continues to foster dialogue between scholars inclined toward the natural sciences and those inclined toward the humanities.

19   The idea of the worldwide diffusion of ideas and methods is rooted in a neo-Kantian perception of history as an independent entity, in romanticist research on language and mythology, and in the German movement of historicism, which is connected closely to Leopold von Ranke (1795–1886). Historicism emerged at a time when the idea of overall progress was increasingly questioned (Streck 2000: 42) and was advocated by the geographers Georg Gerland (1833–1919) and Friedrich Ratzel (1844–1938) in their application of Moritz Wagner's (1813–1887) idea of diffusion. Wagner's fields of interest included natural history, zoology, and geography, as well as ethnography. In 1911, Boas mentioned Ratzel, a zoologist and geographer, as one of his most influential early mentors (Voget 1970: 209).

20   'Die fruchtbarsten Einwände gegen die spekulativen Entwürfe der Evolutionisten kamen im 20. Jahrhundert von den Vertretern einer sich formierenden empirischen Ethnologie: Boas, Kroeber, Lowie, Malinowski, Radcliffe-Brown ...' (Streck 2000: 62). The emerging school of social anthropology in Britain (Malinowski 1915, 1922; Radcliff-Brown 1922; Evans-Pritchard 1937) advocated the method of participant observation in fieldwork, focusing on daily social interaction in a human community rather than on the classification of cultures by the language, artefacts, and physical traits of its representatives. In German-speaking countries, cultural diffusionism took an equally critical stand on the question of cultural evolution through *Kulturkreislehre*, elaborated on by the Vienna School of historical ethnography in accordance with the traditions of German historicism.

21   Instead of cultural evolution in progressive stages, Kulturkreislehre advocated the idea of cultural degeneration, manifested in the supposed decline of monotheism to polytheism (Schmidt 1912–1955) or in the historical development of *Primärkultur* and *Sekundärkultur* in a degenerative process that spoiled features such as monogamy, monotheism, and patriarchal structures, which remained abundant in the assumed *Urkultur* (Rössler 2007: 13). This idea of decay, prominent in fin de siècle thought in the German-speaking countries of central Europe (Spengler 1918, 1922), proved to have a lasting influence on German and Austrian ethnography

and structuralism transpired into the realms of linguistics and cultural and social anthropology in the beginning of the 20th century.

THE 'RENAISSANCE' OF CO-EVOLUTIONARY THEORIES

After the general dismissal of evolutionary theory in biology, linguistics, and cultural anthropology in the beginning of the 20th century, evolutionism made great inroads again. Following the neo-Darwinian 'wedding of [Mendelian] genetics to evolutionary biology' (Hull 1988: 57) in the 'new synthesis' of the 1930s and 1940s (Dobzhansky 1937; Huxley 1942; Mayr 1942), cultural neo-evolutionism was introduced as a controversial theory in American anthropology (Steward 1955; White 1949; Sahlins and Service 1960).[22] In the 1980s, modern population genetics was combined with long-range linguistic comparison (Greenberg 1963, 1971, 1987, 2000/2002) and 'archaeological genetics' (Renfrew 1987; Renfrew and Foster 2006). Linguists, psychologists, population geneticists, and archaeologists have made such attempts to synthesise cultural and linguistic classification with biomapping (e. g. Gray and Atkinson 2003). To a certain degree, these attempts result from the application of mathematic modelling developed in biological genetics to the development of languages and cultures. They can be perceived as the renaissance of an interdisciplinary research agenda in an all-encompassing 'study of human nature', or simply 'anthropology', in the widest sense of the word. I argue that the American tradition of the 'four field approach' perpetuated the anthropological tradition of combining the study of the physical and cultural realms of human existence.

CONCLUSION

In summary, the reconstruction of human biological, linguistic, and cultural co-evolution by the 'new synthesis' of genetic, linguistic, and archaeological data perpetuates three long-standing traditions in the history of the sciences:
(1) the tradition of interdisciplinary intellectual, institutional, and personal contacts among influential scholars;
(2) the tradition of anthropology as an interdisciplinary framework dedicated to the study of the physical and cultural realms of human existence; and

(Völkerkunde) until the 1930s and 1940s (Rössler 2007). Diffusionist ethnography searched for 'pure forms of culture', disregarding the framework of hierarchical development and assuming that major technical and cultural inventions occurred rarely and were transmitted by cultural diffusion rather than by evolution – a view that also prevailed in the works of Boas and his early disciples (Wissler 1926; Kroeber 1939).

22 These representatives of cultural neo-evolutionism did, however, restrict their evolutionary approaches to cultural and social phenomena. A new synthetic approach intending to present a global phylogeny of mankind was introduced in the 1980s, inspired by Richard Dawkins (1976), who coined the word 'meme' to describe cultural replicators.

(3) the temptation provided by the powerful phylogenetic images of the 'tree of life' and the 'tree of languages'.

In reply to the initial question – whether the idea of evolution per se is so convincing or whether our evolutionary thinking is caused mainly caused by consilience in the sense of Edward Wilson (1998) – I have shown that the paradigm of co-evolution was established as a research agenda 200 years ago and has been repeatedly refreshed since then due to reticulations between influential scholars in the humanities and sciences. The theory of co-evolution was generated through lateral transfer between representatives of different scientific realms. Thus, the development of the co-evolutionary paradigm represents a reticulate model of intellectual development, in complete contradiction to the models of unilinear descent that evolutionists in all disciplines favoured.

ACKNOWLEDGEMENTS

The author is greatly indebted to Silvia Fischer and Anja Weigel (both of Ulm University) for final proofreading and to the German Federal Ministry for Education and Research (BMBF) for generously funding the research project 'Evolution and Classification in Biology, Linguistics and the History of the Sciences' from 2009 to 2012.

REFERENCES

Alter, S.G. (1999) Darwinism and the Linguistic Image. Language, Race, and Natural Theology in the Nineteenth Century (Baltimore, London: John Hopkins University Press).

Archibald, J.D. (2009) 'Edward Hitchcock's Pre-Darwinian (1840) 'Tree of Life'', Journal of the History of Biology 42: 561–92.

Atkinson, Q.D. & R.D. Gray (2005) 'Curious Parallels and Curious Connections – Phylogenetic Thinking in Biology and Historical Linguistics', Systematic Biology 54(4): 517.

Augier, A. (1801) Essai d'une nouvelle Classification des Végétaux conforme à l'Ordre que la Nature paroit avoir suivi dans le Règne Végétal: d'ou Resulte une Méthode qui conduit à la Conaissance des Plantes & de leur Rapports naturels. (Lyon: Bruyset Ainé).

Auroux, S. (1990) 'Representation and the place of linguistic change before comparative grammar', in T. de Maro & L. Formigari (eds), Leibniz, Humboldt, and the Origins of Comparativism. Amsterdam Studies in the Theory and History of Linguistic Science Series 3 (Amsterdam: John Benjamins): 231–8.

Bachofen, J.J. (1861) Das Mutterecht (Basel: Beno Schwabe).

Barth, F. (2005) 'Britain and the Commonwealth', in C. Hann (ed) One Discipline, Four Ways: British, German, French, and American Anthropology. The Halle Lectures (Chicago & London: University of Chicago Press): 1–57.

Beer, G. de (1971) 'Darwin, Charles Galton', in C.C. Gillespie (ed) Dictionary of Scientific Biography 3 (New York: Charles Schribers Sons): 563–77.

Bitterli, U. (1991) Die 'Wilden' und die 'Zivilisierten'. Grundzüge einer Geistes- und Kulturgeschichte der europäisch-überseeischen Begegnung (München: Beck).

Boas, F. (1913) Kultur und Rasse (Berlin: Gruyter).

Boas, F. (1940) Race, Language, Culture (New York: MacMillan).

Bock, K. (1955) 'Darwin and Social Theory', Philosophy of Science 22(2): 123–134.

Bowler, P. (1983) Evolution: The History of an Idea (Berkeley, Los Angeles: University of California Press).

Boxhorn, M. Z. (1647a) Bediedinge van de tot noch toe onbekende Afgodinne Nehalennia, over de dusent ende ettelicke hondert Jahren onder het Sandt begraven, dan onlancx ontdeckt op het Strandt van Valcheren in Zeelandt (Leyden: Willem Christiaens van der Boxe).

Boxhorn, M. Z. (1647b) Vraagen voorghestelt ende Opghedraaghen aan de Heer Marcus Zuerius van Boxhorn over de Bediedinge van de tot noch toe onbekende Afgodinne Nehalennia, onlancx by Hem uytgegeven (Leyden: Willem Christiaens van der Boxe).

Boxhorn, M. Z. (1647c) Antwoord van Marcus Zuerius van Boxhorn op de Vraaghen, hem voorgestelt over de Bediedinge van de tot noch toe onbekende Afgodinne Nehalennia, onlancx uytgegeven. In welcke de ghemeine herkomste van der Griecken, Romeinen, ende Duytschen Tale uyt den Scythen duydelijck bewesen, ende verscheiden Oudheden van dese Volckeren grondelijck ontdeckte ende verklaert (Leyden: Willem Christiaens van der Boxe).

Boyd, R. & P.J. Richerson (1985) Culture and the Evolutionary Process (Chicago: University of Chicago Press).

Brues, A.M. (1977) People and Races (New York: MacMillan).

Cassmann, O. (1594) Psychologia anthropologica, sive animae humanae doctrina (Hanau).

Cavalli-Sforza, L.L. & M.W. Feldman (2003) 'The Application of Molecular Genetic Approaches to the Study of Human Evolution', Nature Genetics [suppl.] 33: 266–275.

Cavalli-Sforza, L.L. & M. Seielstad (2001) Genes, Peoples, and Languages (Berkley: University of California Press).

Cavalli-Sforza, L.L.; Piazza, A.; Menozzi, P. & J. Mountain (1988) 'Reconstruction of Human Evolution: Bringing together Genetic, Archaeological, and Linguistic Data', Proceedings of the National Academy of Sciences of the United States of America 85: 6002–6.

Cavalli-Sforza, L.L.; Menozzi, P. & Alberto Piazza (1994) The History and Geography of Human Genes (Princeton, New Jersey: Princeton University Press).

Chambers, R. [anonymously] (1844) Vestiges of the Natural History of Creation (London, Edinburgh: W. and R. Chambers).

Condorcet, M.J.A. de (1795) Esquisse d'un tableau historique des progrès de l'esprit humain (Paris: Agasse).

Currie, T.E., Greenhill, S.J.; Gray, R.; Hasegawa, T. & R. Mace (2010) 'Rise and Fall of Political Complexity in island South-East Asia and the Pacific', Nature 467: 801–804.

Darwin, C. (1859) On the Origin of Species by Means of Natural Selection, or the Preservation of Favoured Races in the Struggle for Life (London: John Murray) [5th ed. 1869].

Dawkins, R. (1976) The Selfish Gene (Oxford: Oxford University Press).

Descartes, R. (1641) Meditationes de prima philosophia (Paris: M. Soly).

Desmond, A. & J.A. Moore (2009) Darwin's Sacred Cause. Race, Slavery and the Quest for Human Origins (Boston/New York: Harcourt).

Dobzhansky, T. (1937) Genetics and the Origin of Species (New York: Columbia University Press).

Driem, G. van (2005) 'Sino-Austronesian vs. Sino-Caucasian, Sino-Bodic vs. Sino-Tibetan, and Tibeto-Burman as default theory', in Y.P. Prasada, Bh.Y. Govinda et al. (eds), Contemporary issues in Nepalese linguistics (Kathmandu: Linguistic Society of Nepal): 305–38.

Dux, G. (2011) Historico-Genetic Theory of Culture. On the Processual Logic of Cultural Change (Bielefeld: Transcript Sociology).

Engels, F. (1884) Der Ursprung der Familie, des Privateigentums und des Staats. Im Anschluß an Lewis H. Morgans Forschungen (Zürich: Schweizerische Volksbuchhandlung).

Evans-Pritchard, E.E. (1937) Witchcraft, Oracles and Magic among the Azande (Oxford: Oxford University Press).

Ferguson, A. (1767) Essay on the History of Civil Society (London, Edinburgh: A. Millar & T. Caddel).

Frobenius, L. (1898) Ursprung der afrikanischen Kulturen (Berlin: Gebrüder Bornträger).

Gallet, F. (1795/1800) 'Arbre genéalogique des langues mortes et vivantes', in T. de Maro & L.

Formigari (eds) Leibniz, Humboldt, and the Origins of Comparativism, Amsterdam Studies in the Theory and History of Linguistic Science Series 3 (Amsterdam: John Benjamins): 229.

Gatterer, J.C. (1775) Abriss der Universalhistorie nach ihrem gesamten Umfange von Erschaffung der Erde bis auf unsere Zeiten erste Hälfte nebst einer vorläufigen Einleitung von der Historie überhaupt und der Universalhistorie insbesonderheit wie auch von den bisher gehörigen Schriftstellern (Göttingen: Vandenhoeck).

Gingrich, A. (2005) 'The German-speaking Countries', in C. Hann (ed) One Discipline, Four Ways: British, German, French, and American Anthropology. The Halle Lectures (Chicago, London: University of Chicago Press): 59–153.

Girtler, R. (1979) Kulturanthropologie (München: Deutscher Taschenbuchverlag).

Gobineau, A. de (1853/1855) L'essai sur l'inégalité des races humaines (Paris: Firmin-Didot).

Graebner, F. (1905) 'Kulturkreise und Kulturschichten in Ozeanien', Zeitschrift für Ethnologie 37: 28–53.

Graebner, F. (1911) Methoden der Ethnologie (Heidelberg: Winter).

Gray, R.D. & Q.D. Atkinson (2003) 'Language-Tree Divergence Times Support the Anatolian Theory of Indo-European Origin', Nature 426: 435–9.

Gray, R.D. (2005) 'Pushing the Time Barrier in the Quest for Language Roots', Science 309: 2007–8.

Greenberg, J.H. (1963) The Languages of Africa (The Hague, Bloomington: Mouton, Indiana University Center).

Greenberg, J.H. (1971) 'The Indo-Pacific Hypothesis', in T.A. Seboek (ed), Linguistics in Oceania. Current Trends in Linguistics 8 (The Hague: Mouton): 807–71.

Greenberg, J.H. (1987) Language in the Americas (Palo Alto: Stanford University Press).

Greenberg, J.H. (2000/2002) Indo-European and its Closest Relatives: The Eurasiatic Language Family, 2 vol. (Palo Alto: Stanford University Press).

Greenhill, S.J.; Blust, R. & R.D. Gray (2008) 'The Austronesian Basic Vocabulary Database: From Bioinformatics to Lexomics', Evolutionary Bioinformatics 4: 271–83.

Gyarmathi, S. (1799) Affinitas linguae Hungaricae cum linguis Fennicae originis grammatice demonstrata (Göttingen: Dieterich).

Haeckel, E. (1874) Anthropogenie oder Entwicklungsgeschichte des Menschen. Gemeinverständliche wissenschaftliche Vorträge über die Grundzüge der menschlichen Keimes- und Stammes-Geschichte (Leipzig: Engelman).

Hann, C. (ed) (2005) One Discipline, Four Ways: British, German, French, and American Anthropology. The Halle Lectures (Chicago, London: University of Chicago Press).

Hoßfeld, U. (2005) Geschichte der biologischen Anthropologie in Deutschland (Stuttgart: Steiner).

Hull, D.L. (1988) Science as a Process. An Evolutionary Account of the Social and Conceptual Development of Science (Chicago, London: University of Chicago Press).

Humboldt, W. von (1820): Über das Vergleichende Sprachstudium in Beziehung auf die verschiedenen Epochen der Sprachentwicklung (Leipzig: Felix Meiner).

Humboldt, W. von (1836) Über die Verschiedenheit des menschlichen Sprachbaus und ihren Einfluß auf die geistige Entwicklung des Menschengeschlechts (Berlin: Königliche Akademie der Wissenschaften, F. Dümmle).

Hutton, C.M. (2005) Race and the Third Reich. Linguistics, Racial Anthropology and Genetics in the Dialectic of Volk (Cambridge: Cambridge University Press).

Huxley, J. (1942) Evolution: The Modern Synthesis (London: Allen & Unwin).

Jakobson, R. (1931) 'Über die phonologischen Sprachbünde', in R. Jakobson (ed), Selected Writings 1: Phonological Studies (The Hague: Nijhoff): 137–43.

Jones, W. (1786) 'The Third Anniversary Discourse, on the Hindus, delivered by the President, February 2, 1786', Asiatic Researches 1: 415–31.

Koerner, K. (1983) Linguistics and Evolutionary Theory. Three Essays by August Schleicher, Ernst Haeckel, and Wilhelm Bleek (Amsterdam, Philadelphia: John Benjamins).

Koppers, W. (1915–16) 'Die ethnologische Wirtschaftsforschung: Eine historisch-kritische Studie', Anthropos 10: 611–51, 11: 971–1079.

Kressing, F. (2012) 'Screening Indigenous Peoples' Genes – The End of Racism or Postmodern Bio-Imperialism?', in S. Berthier; S. Tolazzi; S. Whittick. (eds) Biomapping or Biocolonizing? Indigenous Identities and Scientific Research in the 21st Century (Amsterdam, New York: Rodopi): 117–36.

Kressing, F.; Fangerau, H. & M. Krischel (2013) 'The 'Global Phylogeny' and its Historical Legacy – A Critical Review of a Unified Theory of Human Biological and Linguistic Co-Evolution', Medicine Studies, February (online), DOI 10.1007/s12376_013_0081-8.

Krischel, M.; Kressing, F. & H. Fangerau (2011) 'Netzwerke statt Stammbäume in der Wissenschaft? Die Entwicklung der evolutionären Theorie als wechselseitiger Transfer zwischen Geistes- und Naturwissenschaften', in M. Krischel & H.K. Keul (eds) Deszendenztheorie und Darwinismus in den Wissenschaften vom Menschen (Stuttgart: Steiner): 107–21.

Kroeber, A.L. (1939) Cultural and Natural Areas of Native North America (Berkeley: University of California Press).

Lamarck, J.-B. (1809) Philosophie zoologique, ou, exposition des considérations relative à l'histoire naturelle des animaux (Paris: Dentu).

Lapouge, G.V. (1899) L'aryen et son rôle social (Paris: Librairie Payot).

LeBon, G. (1894) Le lois psychologiques de l'évolution des peuples (Paris: Félix Alcan).

Leclerc, G. (1972) Anthropologie et colonialisme (Paris: Fayard).

Malinowski, B. (1915) The Trobriand Islands (London: Routledge).

Malinowski, B. (1922) 'Argonauts of the Western Pacific: An Account of Native Enterprise and Adventure in the Archipielagoes of Melanesian New Guinea', Studies in Economics and Political Science 65 (London: Routledge & Kegan Paul).

Marks, J. (1995) Human Biodiversity. Genes, Race, and History (New York: Aldine de Gruyter).

Mayr, E. (1942) Systematics and the Origin of Species (NewYork: Columbia University Press).

McMahon, A. & R. (2010) 'Genetics, Historical Linguistics and Language Variation', Language and Linguistics Compass 2(2): 264–88.

McMahon, R. (2004) 'Genes and Languages', Community Genetics 7: 1–13.

Meiners, C. (1785) Grundriß der Geschichte der Menschheit (Lemgo: Meyer).

Müller, M.F. (1855) Languages of the Seat of War in the East, with a Survey of the Three Families of Language, Semitic, Arian, and Turanian (London: Williams & Norgate).

Müller, M.F. (1861) Lectures on the Science of Language (London: Royal Institute).

Parkin, R. (2005) 'The French-Speaking Countries', in C. Hann (ed), One Discipline, Four Ways: British, German, French, and American Anthropology. The Halle Lectures (Chicago, London: University of Chicago Press): 157–253.

Radcliff-Brown, A.R. (1922) The Andaman Islanders. A Study in Social Anthropology (London: Cambridge University Press).

Ragan, M.A. (2009) 'Trees and Networks before and after Darwin', Biology Direct 4: 1–38.

Renfrew, C. & P. Forster (2006) Phylogenetic Methods and the Prehistory of Languages (Cambridge: McDonald Institute for Archaeological Research).

Renfrew, C. (1987) Archaeology and Language: The Puzzle of Indo-European Origins (Cambridge: Cambridge University Press).

Römer, R. (1989) Sprachwissenschaft und Rassenideologie in Deutschland (München: Fink).

Rössler, M. (2007) 'Die deutschsprachige Ethnologie bis ca. 1960: Ein historischer Abriss', Cologne Working Papers in Cultural and Social Anthropology 1, http://kups.ub.uni-koeln.de/volltexte/2007/1998/pdf/kae0001.pdf, accessed March 11, 2013.

Sahlins, M.D. & E.R. Service (1960) Evolution and Culture (Ann Arbor: University of Michigan Press).

Sahlins, M.D. (2000) 'Ethnographic Experience and Sentimental Pessimism: Why Culture is not a Disappearing Object', in L. Daston (ed), Biographies of Scientific Objects (Chicago, London: University of Chicago Press): 158–202.

Sapir, E. (1921) Language. An Introduction to the Study of Speech (New York: Harcourt Brace).

Schlegel, F. (1808) Ueber die Sprache und Weisheit der Indier. Ein Beitrag zur Begründung der Alterthumskunde (Heidelberg: Mohr & Zimmer).

Schleicher, A. (1853) 'Die ersten Spaltungen des indogermanischen Urvolkes', Allgemeine Monatsschrift für Wissenschaft und Literatur: 786–7 [101–2].

Schleicher, A. (1861/1862) Compendium der vergleichenden Grammatik der indogermanischen Sprachen. Kurzer Abriss der indogermanischen Ursprache, des Altindischen, Altiranischen, Altgriechischen, Altitalischen, Altkeltischen, Altslawischen, Litauischen und Altdeutschen, 2 vol. (Weimar: Böhlau).

Schleicher, A. (1863) Die Darwinsche Theorie und die Sprachwissenschaft. Offenes Sendschreiben an Herrn Dr. Ernst Haeckel, o. Professor der Zoologie und Director des zoologischen Museums an der Universität Jena (Weimer: Böhlau).

Schlözer, A.L. (1772) Vorstellung einer Universal-Historie (Göttingen, Gotha: Dieterich).

Schmidt, W. (1912–1955) Der Ursprung der Gottesidee. Eine historisch-kritische und positive Studie, 12 vol. (Münster: Aschendorff).

Schwidetzky, I. (1992) History of biological anthropology in Germany. Occasional Papers 3/4 (Newcastle, Tyne: International Association of Human Biologists).

Spencer, H. (1857) Progress: It's Law and Cause (London: Williams & Norgate).

Spengler, O. (1918) Der Untergang des Abendlandes. Gestalt und Wirklichkeit (Wien: Braumüller).

Spengler, O. (1922) Der Untergang des Abendlandes. Welthistorische Perspektiven (München: Beck).

Stagl, J. (1974) Kulturanthropologie und Gesellschaft. Wege zu einer Wissenschaft (München: List).

Stevens, P.F. (1983), 'August Augier's Arbre Botanique (1801), A Remarkable Early Botanical Representation of the Natural System', Taxon 32: 203–11.

Steward, J. (1955) Theory of Culture Change. The Methodology of Multilinear Evolution (Urbana: University of Illinois Press).

Stocking, G.W. (1988) Bones, Bodies, Behavior. Essays on Biological Anthropology (Madison, London: University of Wisconsin Press).

Streck, B. (2000) 'Kulturanthropologie', in B. Streck (ed) Wörterbuch der Völkerkunde (Wuppertal: Hammer Verlag, Edition Trickster): 141–4.

Suttrop, U. (1999) 'Diskussionsbeiträge zur Stammbaumtheorie', in Ago Künnap (ed) Fenno-Ugristica 22: 223–51.

Trigger, B.G. (1998) Sociocultural Evolution: Calculation and Contingency (Oxford: Blackwell).

Uschmann, G. (1972) 'Haeckel, Ernst Heinrich Philipp August', in C.C. Gillespie (ed), Dictionary of Scientific Biography 6 (New York: Charles Schribers Sons): 6–11.

Voget, F.W. (1970) 'Franz Boas', in C. Ch. Gillispie (ed) Dictionary of Scientific Biography, 1 (New York: Ch. Schribners Sons): 207–13.

White, L. (1949) The Science of Culture: A Study of Man and Civilization (New York: Farrar, Straus and Giroux).

Wilson, E.O. (1998) Consilience: The Unity of Knowledge (New York: Alfred L. Knopf)

Wissler, C. (1926) The Relation of Nature to Man in Aboriginal America (New York: Oxford University Press).

Witsen, N. (1692) Noord en Oost Tartarye, ofte Bonding Ontwerp van eenige dier Landen en Volken, welke voormaels bekent zijn geweesst …, 2 vol. (Amsterdam: F. Halma).

Zittel, C. (2009) Theatrum philosophicum. Descartes und die Rolle ästhetischer Formen in der Wissenschaft (Berlin: Akademie-Verlag).

2. PHYLOGENETIC CLASSIFICATIONS AND NETWORK APPROACHES IN LINGUISTICS AND BIOLOGY

DO LANGUAGES GROW ON TREES?
THE TREE METAPHOR IN THE HISTORY OF LINGUISTICS

Hans Geisler and Johann-Mattis List

THE RETURN OF THE TREES

Among biologist as well as linguists, it is now widely accepted that there are many striking parallels between the evolution of life forms and the history of languages. Starting from the rise of language studies as a scientific discipline in the early 19th century up to today's recent "quantitative turn" in historical linguistics, scholars from both disciplines have repeatedly pointed to similarities between the respective research objects in biology and linguistics. Of all these parallels, the use of *family trees* to model the differentiation of species (genomes and languages) is surely the most striking one. Methodically speaking, genealogical relations between languages and species can both be visualized with help of bifurcating trees which indicate the splitting of ancestral into descendant taxa. Being developed independently in linguistics and biology (Hoenigswald 1963), the tree model suffered different fates in both disciplines: While the reconstruction of phylogenetic trees successively became one of the key objectives in evolutionary biology, the tree model was controversially disputed in linguistics and – although to no time completely abandoned – never became a true part of the consensus.

Although linguists always had certain reservations regarding the tree model, it recently experienced a surprising revival. While earlier linguistic work on phylogenetic reconstruction was almost exclusively based on the intuitive weighting of features from very small samples of well-studied ancient languages, the integration of stochastic methods originally designed for biological applications made it possible to analyze large quantitative datasets automatically (Gray and Atkinson 2003; Atkinson and Gray 2006). Whereas tree construction has played for some time a minor role in historical linguistics, it has again become a specific field of historical linguistic endeavor in the last two decades (Pagel 2009: 414).

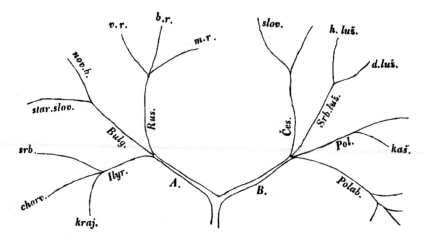

Figure 1: An early family tree of the slavic languages by Čelakovský (1853)

In the following, we will give a short overview how the tree model emerged and made his way into 19th and 20th century linguistics. Though nearly ousted by competing alternative models like the wave theory for quite a long time, bifurcating trees has prevailed recently because they make it possible to represent logically dichotomous relations between species and languages which in turn allow for easy phylogenetic algorithmization. Unfortunately, the tree model's logical simplicity masks the complexity of biological and linguistic research objects in many respects. The most important surely is the wide neglect of language contact and ensuing borrowing. Thus, we claim that only combined approaches which describe both the vertical and the horizontal components of language relations are apt to depict the intrinsically distorted character of language change adequately.

THE ORIGIN OF THE TREES

According to the current view in historical linguistics, one can roughly distinguish two different kinds of language relations: *genealogical language relations*, i.e. relations which are due to the common descent from an ancestor language, and *non-genealogical language relations*, i.e. relations which are a result of language contact. One of the key tasks of historical linguistic research is to find out whether resemblances between languages are a result of the former or the latter kind of language relationship. Otherwise, no language history could be drawn.

To infer whether specific resemblances between languages are due to contact or due to inheritance, however, is a complicated task, and in many cases there is no clear-cut procedure to discriminate between the two. The deeper one goes back in time, the greater becomes the problem of inference. Thus, German *Kopf* "head" and English *cup* "cup" probably go back to Proto-Germanic **kuppa-* "cup" (Orel 2003), yet whether the word was borrowed from Latin into Proto-Germanic (Kluge and

Seebold 2002) or inherited from Proto-Indo-European (Orel 2003) cannot be re-solved with full confidence (see Figure 2). It is therefore not surprising that the idea that language relations can be divided into genealogical and non-genealogical ones was developed considerably late. Before the 19th century, the dominant view on language relations was non-genealogical. Discussions regarding the origination of languages were restricted to the biblical myth of the *Tower of Babel*.

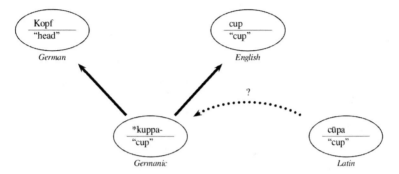

Figure 2: German Kopf and english cup: inheritance or borrowing?

Early views on language relations

Catastrophism as opposed to *gradualism* (or *evolutionarism*) was a leading para-digm in scholarly thinking up to the early 19th century. While in fields like geology and biology the biblical creation myth accounted for the origin of the earth and the species (Christy 1983), the origin of all languages was explained by the biblical myth of the *Tower of Babel* according to which all languages originated as aberra-tions of a single language after the *confusion of tongues*. As in geology and biology the origin of a given diversity was explained as the result of an ad-hoc *catastrophic* event. All languages where assumed to be derived from the mysterious "Adamic Language" which became later directly identified with Hebrew. Consequently, the *Hebrew Paradigm* heavily influenced the way scholars would investigate language relations (Klein 2004). Since the monophyletic origin of all languages was appar-ently already proven by the religious dictum and linguistic change was character-ized as an abrupt process of decay, the scholars mostly restricted their research to speculative etymological studies trying to show that all languages had inherited at least some words of Hebrew (cf., e.g., the work of Münster 1523, Reuchlin 1506, or Cruciger 1616).

While – held back by the Hebrew Paradigm – the genealogical perspective was only sporadically adopted and investigated. Scholars were well aware of the fact that languages can influence each other in many different ways. The non-genealog-ical perspective on language relations was the prevailing one in pre-19th century linguistics (Allen 1953: 55–7), and there are many examples in the literature, where scholars explicitly make use of non-genealogical explanations in order to explain specific resemblances between certain languages (see, e.g., Cratylus, Institutio Oratoria, Webb 1787). This is obviously due to the fact that language contact is fairly easy to recognize, not only for those who show a special interest in languages but also for "normal" speakers who are in contact to people who speak in different tongues.

The discovery of tree-likeness

August Schleicher (1821–1868) is often regarded as the founding father of historical linguistics, being the one who established it as a real science (Fox 1995: 23–7). His two main contributions to historical linguistics were the method of linguistic reconstruction (Schleicher 1861) and the development of the tree model to visualize genealogical language relations.

Schleicher's tree model (Schleicher 1853a and 1853b) is the cumulation of several findings which were made during the early 19th century. In this time, scholars such as Jacob Grimm (1785–1863) and Rasmus Rask (1787–1832), had detected that – in contrast to previous opinions – certain aspects of languages, namely their sound systems, did not change *chaotically*, but *regularly*, making it possible to compare different languages systematically for common traits (see Rask 1818, Grimm 1822). Along with Franz Bopp's (1791–1867) independent detection of many grammatical resemblances between Sanskrit and many European languages such as Latin, Greek, and Gothic (see Bopp 1816), it seemed, for the first time, no longer possible to explain these similarities by accident or derivation, but only by a common origin of these specific languages. Abandoning the Hebrew Paradigm, and adopting the hypothesis that sound change was a regular process, scholars apparently had finally found a method by which it was possible to distinguish vertical from horizontal language relations.

The *regularity hypothesis* upon which the new vertical thinking in linguistics was built summarizes three major characteristics of sound change which are already explicitly mentioned in Schleicher's early work. According to these characteristics, sound change is a *universal*, a *gradual*, and a *law-like* process (cf. Schleicher 1848: 25). *Universality* implies that the process is independent of time and space, *graduality* implies that the process is neither abrupt nor chaotic, and *law-likeness* implies that the process is (to a great degree) exceptionless.

Perspective	Hebrew paradigm	Schleicher
heredity	sporadic	systematic
differentiation	singular	recurrent
change	chaotic	regular

Table 1: The hebrew paradigm and Schleicher's 'Tree Model'

The new theory of vertical language relations which is directly reflected in the tree model, radically differs from the earlier conception of language relations within the Hebrew Paradigm: *change* is no longer seen as a *chaotic*, but as a *regular* process, *heredity* is no longer believed to be a *sporadic* but a *systematic* phenomenon, and the *origination* of new languages is no longer identified with a *singular* event but as a process which repeatedly occurs during all times (see Table 1).

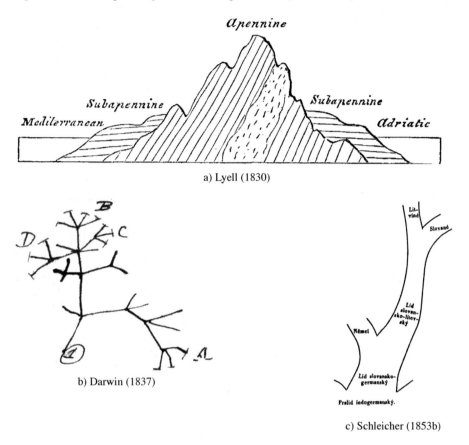

a) Lyell (1830)

b) Darwin (1837)

c) Schleicher (1853b)

Figure 3: Illustrations by Lyell (a), Darwin (b), and Schleicher (c)

Common paradigms in Geology, Biology and Linguistics

At about the same time when linguists realized that the apparently chaotic and sporadic phenomena of language change where indeed universal and gradual, geologists and biologists came to similar conclusions in their own fields. Between 1830 and 1833 the English geologists Charles Lyell (1797–1875) published his multivolume book *Principles of Geology* (Lyell 1830–1833) in which he substantiated the claim – first brought forward by James Hutton (1726–1797) – that the shape of the earth was the result of slow-moving and gradually operating forces which were acting independently of times and places. In 1859 the English biologist Charles

Darwin (1809–1882) published the famous book *On the Origin of Species* in which he first introduced the idea that the diversity of life was due to the universally and gradually operating force of natural selection (Darwin 1859). Whether these new ideas regarding the universality and graduality of certain processes in different disciplines were due to mutual influence or due to the spirit of the age: The new paradigm of *uniformitarianism* made it possible to reconstruct the prehistory of regions, species, and languages under the common slogan "The present is a key to the past", and scholars from all three disciplines noticed and discussed the nature and the implication of the parallels they found in the different fields of research.

Back to Dendrophobia

In contrast to evolutionary biology, where family trees became the leading paradigm for the description of species differentiation, the popularity of language trees soon began to fade in the newly established discipline of historical-comparative linguistics. In 1872 Johannes Schmidt (1843–1901) published the book *Die Verwandtschaftsverhältnisse der indogermanischen Sprachen* (Schmidt 1872) in which he pointed to various problems regarding the applicability and the adequacy of the tree model. He pointed out that the data of the Indo-European languages did not suggest a simple tree-like differentiation. In order to account for his findings, he proposed the so-called *Wave Theory* according to which certain changes spread like waves in concentric circles over neighboring speech communities. Even two years earlier Hugo Schuchardt (1842–1927) criticized the assumption that languages simply split and then evolve independently (as suggested by the tree model), emphasizing that languages usually diverge gradually while at the same time mutually influencing each other: 'We connect the branches and twigs of the tree with countless horizontal lines and it ceases to be a tree' (Schuchardt 1870 [1900] 11)[1].

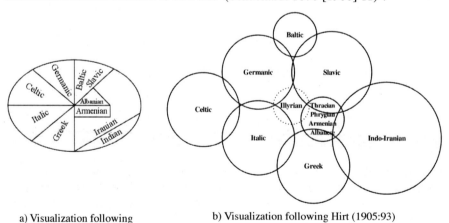

a) Visualization following   b) Visualization following Hirt (1905:93)
   Meillet (1908:134)

---

1   Translation of the authors, original text: 'Wir verbinden die Äste und Zweige des Stammbaums durch zahllose horizontale Linien, und er hört auf ein Stammbaum zu sein.'

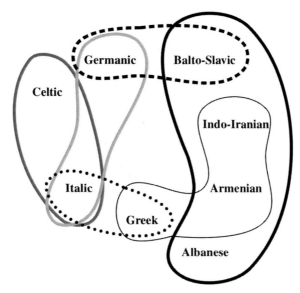

c) Visualization following Bloomfield (1933:316)

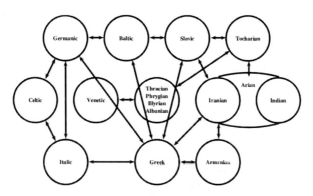

d) Visualization following Bonfante (1931:174)

Figure 4: The 'Wave Theory' in different visualizations

While most of the scholars were well aware of the inadequacy of the family-tree model, they had great difficulties in coming up with a conclusive alternative model which would describe and depict the complex reality of the phylogenetic history of languages in an equally simple and straightforward way. The fruitless quest for new metaphors is reflected in numerous different visualizations of Schmidt's Wave Theory ranging from simple geographical maps (Schmidt 1875: 199; Meillet 1908: 134), via overlapping circles (Hirt 1905: 93) or alternating boundaries (Bloomfield 1933: 316), up to networks (Bonfante 1931: 174), as illustrated in Figure 5.

What all these visualizations have in common is that they emphasize the spatial extension of languages which is neglected within the tree-model. At the same time, however, the time dimension is sacrificed: Languages are arranged on a map and relations between the languages are marked, yet all relations are displayed as static differences, not as dynamic processes of differentiation. Surely, this lack of dynamicity was one of the reasons, why linguists never abandoned the family tree completely, but rather used both models in dependence of the respective problems they were dealing with.

PROBLEMATIC ASPECTS OF THE 'TREE MODEL'

The tree model can be criticized by questioning its practicability, its plausibility, or its adequacy. Although criticism regarding plausibility and adequacy seems to be stronger than criticism regarding practicability, most of the arguments which have been brought forward against the tree model belong to the latter kind.

Practicability of the model

Many of the early opponents of Schleicher's *Stammbaum* disfavored the tree model because they experienced problems when trying to apply it. Most of these cases were due to conflicts in the data: Apparently, the tree model could not account for the distribution of common features in the descendant languages. Thus, applying a quasi-quantitative account, Schmidt (1872) listed words which were patchily distributed over the major Indo-European subgroups in support for his Wave Theory. In his counts, for example, there are 132 words which are reflected in both Latin and Old Greek but not in Old Indian, 99 words which are reflected in both Old Indian and Old Greek but not in Latin, yet only 20 words which occur in Latin and Old Indian but not in Old Greek. Following Schmidt's line of thought, these counts contradict the tree model, since they suggest a strange pattern of closeness between the three languages where Old Greek is close to both Latin and Old Indian while Old Indian and Latin are only close to Old Greek (see Figure 5a).

However, this argumentation has a striking shortcoming, in so far as it ignores the *temporary status* of knowledge in the historical sciences. Thus, Schmidt's estimations for common roots between Latin and Old Indian are considerably low. According to estimates drawn from Nicolaev (2007), there are 364 cognate words in Old Greek and Old Indian which are not reflected in Latin, 199 between Old Indian and Latin which are not reflected in Old Greek, and 379 between Latin and Old Greek which are not reflected in Old Indian (see Figure 5b). This shows that drawing conclusions from historical data is always preliminary. If a current state of knowledge disfavors the tree model, this doesn't need to hold for future states.

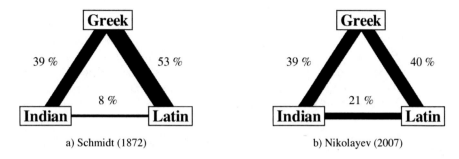

Figure 5: Cognate counts in old Greek, Latin, and old Indian

A further point which is mentioned by Schmidt himself is the impact of hidden bor-rowings. As it was mentioned before, the deeper one goes back in time, the more difficult it becomes to distinguish clearly between similarities due to genealogical and similarities due to non-genealogical relations. If the high amount of shared cognates between Old Greek and Latin turned out to be the result of the close con-tact between the two languages, this would, however, not contradict the tree model, it would only show that it is of crucial importance to disentangle vertical and hori-zontal relations before the reconstruction of family trees can be applied faithfully.

Plausibility of the model

From the above-mentioned one can conclude that pointing to the impracticability of the tree model cannot consistently prove its inadequacy, since the practicability of a given model can be overcome by advanced methods. Objections regarding plau-sibility, on the other hand, are much stronger, since they seriously challenge the tree model.

The arguments raised by the opponents of the tree model come along with its obvious simplifications: When mapped onto a family tree, languages are reified and treated as discrete objects located in space and time. Language divergence is neces-sarily characterized as an abrupt event, and no reverse process of convergence is allowed. In reality such a situation is met only under rare circumstances when the speakers of a language separate geographically. Under normal circumstances, how-ever, languages form areal continua of slightly diverging varieties. A strict separa-tion of languages does only hold for distantly related, standardized, written lan-guages. Lacking the geographical dimension, the family tree can neither model language divergence in all its complexity, nor can it account for the opposite pro-cess of convergence which eventually may even lead to hybridization. Thus, from what is known from studies on dialect geography and language divergence, there are obvious plausibility issues when trying to model language history with the help of trees only.

Adequacy of the model

Apart from the apparent plausibility issues arising from Schleicher's Stammbaum, the discontent of most linguists with family trees surely also results from their lack of adequacy. If the goal of historical linguistics is to describe realistically how languages evolve, it is surely not enough to simply point to their vertical history, since the horizontal aspects of language history are surely at least as – if not even more – characteristic for language history as the vertical ones. The tree model's lack of expressiveness is surely one of the most important reasons for the general reluctance of linguists to draw phylogenetic trees: If family-trees are simply not realistic enough to depict what linguists know about the history of the languages they investigate, why should one even make the effort to reconstruct them?

SPECIES EVOLUTION AND LANGUAGE CHANGE

The key assumption of the new approaches in historical linguistics is that the characteristic processes of language change and biological evolution are so similar that the methods designed for one discipline may also be used in the other one, despite the fact that the domains differ (Croft 2008: 225). The use of biological methods requires certain analogies to be made between linguistic and biological processes and entities. Table 2 lists some of the most common ones which can often be found in the literature. Thus, regarding the *unit of heredity*, the biological *gene* is usually set in analogy with the linguistic *word*, both being 'discrete heritable units' (Pagel 2009: 406). *Replication* of the heritable units is achieved via concrete mechanisms of *reproduction* in biological evolution and via *learning* in language history. From the perspective of *origination*, *cladogenesis* in biology is identified with *language splitting* in linguistics (ibid.). From the perspective of *change*, the driving forces of biological evolution such as *natural selection* and *genetic drift* are compared with *social selection* and *trends* eventually leading to language change (ibid.). Last not least, differentiation is usually assumed to be treelike, and the impact of "horizontal forces" on evolution is considered to be rather low in both cases.

Aspects	Species	Languages
unit of heredity	gene	word
replication	(asexual and sexual) reproduction	learning
origination	cladogenesis	language splitting
forces of change	natural selection and genetic drift	social selection, trends
differentiation	treelike	treelike (?)

Table 2: Some apparent parallels between species and languages

Assuming that these parallels hold, it seems perfectly plausible to use the methods developed for the application in one discipline in the other. However, it is important to be aware not only of the parallels but also of the differences between the research objects of both disciplines. The most striking difference between languages and genomes is that biological evolution manifests itself *substantially* while language history does not. In terms of Popper (1978), genome evolution and language evolution take place in different *worlds*: While biological organisms are part of *world 1*, the 'world that consists of physical bodies' (ibid. 143), languages belong to *world 3*, the 'world of the products of the human mind, such as languages; tales and stories and religious myths' (ibid. 144) which are replicated by learning.

Since we are dealing with very different domains here, the processes dominating in biological and linguistic evolution may also differ quite significantly. Thus, the *unit of heredity* in biology, the gene, is built from a set of universal characters which can be found in all organisms. The *unit of heredity* in linguistics, the word, however, is built from a set of sounds which are distinctive only with respect to the language they belong to. Unlike genes, words are not drawn from a universal alphabet, but from alphabets which themselves are subject to change. Therefore, biological methods which only work on global similarity, such as traditional alignment algorithms, necessarily fail to detect these specific similarities which are of interest to historical linguistic.

When using biological methods in linguistic applications, it is therefore important to be very cautious, and to check whether the parallels really hold or whether one is simply led astray by some first-glance lookalikes. It seems that the use of biological methods in historical linguistics is not always based on a thorough reflection regarding the question of comparability. Phylogenetic reconstruction, for example, is usually based on cognate-sets extracted from lexicostatistical wordlists. In these analyses, the wordlists, reflecting the so-called *basic vocabulary* of the languages under investigation (Swadesh 1955), are usually compared with the *core genome* in biology, i. e. they are supposed to represent the most stable, slow-changing, and least borrowing-prone part of a language's lexicon. However, given the independence of word form and meaning, which does *not* hold for biology, there is no objective procedure to determine a language's basic items. As a result, the creation of basic lists in linguistics is based on a manual procedure which could be shown to be very prone to errors in item translation (Geisler & List 2013) and the identification of borrowings (Nelson-Sathi et al. 2011).

ALTERNATIVES TO THE 'TREE MODEL'

Given that the family tree is not sufficient to model language history in all its complexity, while the Wave Theory lacks the dynamicity of the tree model, remaining a mere static, map-like visualization of shared similarities, one may raise the question whether there are any other possibilities to display both genealogical and non-genealogical relationships between languages. Given that both tree and wave reflect certain aspects of language relations, the most straightforward alternative would be to combine both models in a network approach where both horizontal and vertical language relations are displayed (see Nelson-Sathi et al. this volume). Such an approach preserves the advantage of the tree model's dichotomous logic with clear-cut categorizations, but further allows fine-graded mapping of language contact. Although the idea of combining trees and waves has been developed very early in the history of linguistics, there are only a few attempts to visualize or formalize it (Southworth 1964; Holzer 1995), and it was only recently that a quantitative approach for the reconstruction of phylogenetic networks based on lexicostatistical wordlists has been proposed (Nelson-Sathi et al. 2011). Nevertheless, given the complexity of language history, combined networks of horizontal and vertical language relations seem to offer a promising alternative to both trees and waves in historical linguistics.

CONCLUSION

In this paper we showed that during the history of linguistics the family tree never played a major role. Soon after the model was first introduced, scholars criticized the concept for its obvious shortcomings and proposed various other ways to model language history, none of which gained broad acceptance. The recent quantitative turn in historical linguistics which was initiated by the adaptation of new automatic methods initially designed for evolutionary biology led to an unexpected revival of the tree model in historical linguistics. Although the new methods doubtlessly decrease the amount of subjectivity inherent in the traditional intuitive approaches to phylogenetic reconstruction, they do not cope for the simplifying character of the tree model *per se*. In order to model language history in a realistic way, combining approaches which reflect the vertical as well as the horizontal aspects of language relations are needed.

REFERENCES

Allen, W.S. (1953) 'Relationship in comparative linguistics', in Transactions of the Phylological Society, 52–108.

Atkinson, Q.D. & R.D. Gray (2006) 'How old is the Indo-European language family? Illumination or more moths to the flame?' in P. Forster and C. Renfrew (eds), Phylogenetic methods and the prehistory of languages. McDonald Institute monographs (Cambridge/Oxford/Oakville: McDonald Institute for Archaeological Research): 91–109.

Bloomfield, L. (1933 [1973]) Language (London: Allen & Unwin).

Bonfante, G. (1931) 'I dialetti indoeuropei', in Annali del R. Istituto Orientale di Napoli 4: 69–185.

Bopp, F. (1816) Über das Conjugationssystem der Sanskritsprache in Vergleichung mit jenem der griechischen, lateinischen, persischen und germanischen Sprache. Nebst Episoden des Ramajan und Mahabharas in genauen metrischen Uebersetzungen aus dem Originaltexte und einigen Abschnitten aus den Veda's (Frankfurt am Main: Andreäische Buchhandlung).

Čelakovský, F.L. (1853) Čtení o srovnavací mluvnici slovanské (Prague: Universitě pražské).

Christy, C. (1983) Uniformitarianism in linguistics. Studies in the history of linguistics (Amsterdam/ Philadelphia: John Benjamins).

Cratylus: Plato (ca. 399 BC), English Translation, in Fowler, H.N., Plato in Twelve Volumes (London: William Heinemann).

Croft, W. (2008) 'Evolutionary Linguistics', Annual Review of Anthropology 37, 219–34.

Cruciger, G. (1616) Harmonia linguarum quatuor cardinalium: Hebraicae Graecae Latinae & Germanicae: In Qua Praeter Summum Earum Consensum, acceptionumque propriarum ab impropriis distinctionem, perpetua unius ab altera, origo perspicue deducitur (Frankfurt, Tampach & Bringer).

Darwin, C. (1837) Notebook on Transmutation of Species.

Darwin, C. (1859) On the origin of species by means of natural selection, or, the preservation of favoured races in the struggle for life (London: John Murray).

Fox, A. (1995) Linguistic reconstruction. An introduction to theory and method (Oxford: University Press).

Geisler, H. & J.-M. List (2013) 'Beautiful trees on unstable ground. Notes on the data problem in lexicostatistics', in H. Hettrich (ed) Die Ausbreitung des Indogermanischen. Thesen aus Sprachwissenschaft, Archäologie und Genetik (Wiesbaden: Reichert).

Gray, R.D. & Q.D. Atkinson (2003) 'Language-tree divergence times support the Anatolian theory of Indo-European origin', Nature 426.6965: 435–9.

Grimm, J. (1822) Deutsche Grammatik. 2nd ed. Vol. 1 (Göttingen: Dieterichsche Buchhandlung).

Hoenigswald, H.M. (1963) 'On the History of the Comparative Method', Anthropological Linguistics 5.1: 1–11.

Holzer, G. (1995) Das Erschließen unbelegter Sprachen. Zu den theoretischen Grundlagen der genetischen Linguistik (Frankfurt am Main: Lang).

Hirt, H. (1905) Die Indogermanen. Ihre Verbreitung, ihre Urheimat und ihre Kultur. Vol. 1 (Strassburg: Trübner).

Institutio Oratoria: Quintilian (ca. 50 AD) Institutio Oratoria. English Translation: I S. Watson, (Iowa: Institutes of Oratory).

Klein, W.P. (2004) 'Was wurde aus den Wörtern der hebräischen Ursprache? Zur Entstehung der komparativen Linguistik aus dem Geist etymologischer Spekulation', in G. Veltri & G. Necker (eds), Gottes Sprache in der philologischen Werkstatt. Hebraistik vom 15. bis zum 19. Jahrhundert, Proceedings of the Symposium "Die Geburt der Philologie aus dem Geist der Hebraistik" (Wittenberg, Oct. 6–6, 2002), Studies in European Judaism 11 (Leiden: Brill): 3–23.

Kluge, F. (ed) (2002) Etymologisches Wörterbuch der deutschen Sprache. Cont. by E. Seebold. 24th ed. (Berlin: de Gruyter).

Lyell, C. (1830) Principles of geology, being an attempt to explain the former changes of the Earth's surface, by reference to causes now in operation. Vol. 1. (London: John Murray).

Meillet, A. (1922 [1908]) Les dialectes Indo-Européens (Paris: Librairie Ancienne Honoré Champion).

Münster, S. (1523) Dictionarium Hebraicum, nunc primum editum et typis excusum, adiectis Chaldaicis vocabulis non parum multis. (Basileae: Apud Frobeniu).

Nelson-Sathi, S.; List, J.-M.; Geisler, H.; Fangerau, H.; Gray, R.D.; Martin, W. & T. Dagan (2011) 'Networks uncover hidden lexical borrowing in Indo-European language evolution', Proceedings of the Royal Society B. 1713.278: 1794–803.

Nikolayev, S. (2007) Indo-European etymology. available online: http://starling.rinet.ru/cgi-bin/main.cgi.

Orel, V. (2003) A handbook of Germanic etymology (Leiden: Brill).

Pagel, M. (2009) 'Human language as a culturally transmitted replicator', Nature Reviews 10: 405–15.

Popper, K.R. (1978) 'Three worlds', The Tanner Lectures on Human Values: 143–67.

Rask, R.K. (1818) Undersögelse om det gamle Nordiske eller Islandske sprogs oprindelse (Copenhagen: Gyldendalske Boghandlings Forlag).

Reuchlin, J. (1506) Ioannis Reuchlin Phorcensis LL. Doc. ad Dionysium fratrem suum Germanum de rudimentis Hebraicis libri III (Pforzheim: Thomas Anshelm).

Schleicher, A. (1848) Zur vergleichenden Sprachengeschichte (Bonn: König).

Schleicher, A. (1853a) 'Die ersten Spaltungen des indogermanischen Urvolkes', Allgemeine Monatsschrift für Wissenschaft und Literatur: 786–7.

Schleicher, A. (1853b) 'O jazyku litevském, zvláště na slovanský. Čteno v posezení sekcí filologické král. České Společnosti Nauk dne 6. června 1853', Časopis Čsekého Museum 27: 320–34.

Schleicher, A. (1861) Compendium der vergleichenden Grammatik der indogermanischen Sprache. Vol. 1: Kurzer Abriss einer Lautlehre der indogermanischen Ursprache, des Altindischen (Sanskrit), Alteranischen (Altbaktrischen), Altgriechischen, Altitalischen (Lateinischen, Umbrischen, Oskischen), Altkeltischen (Altirischen), Altslawischen (Altbulgarischen), Litauischen und Altdeutschen (Gotischen) (Weimar: Böhlau).

Schmidt, J. (1872) Die Verwantschaftsverhältnisse [sic!] der indogermanischen Sprachen (Weimar: Böhlau).

Schmidt, J. (1875) Zur Geschichte des indogermanischen Vocalismus (Weimar: Böhlau).

Schuchardt, H. (1900[1870]) 'Über die Klassifikation der romanischen Mundarten', Probe-Vorlesung, gehalten zu Leipzig am 30. April 1870 (Graz: Brevier): 166–88.

Swadesh, M. (1955) 'Towards greater accuracy in linguistic dating', International Journal of American Linguistics 21.1: 121–37.

Webb, D. (1787) 'Some reasons for thinking, that the Greek language was borrowed from the Chinese', in Fourmont, Notes on the Grammatica Sinica of Mons (London: Dodsley).

LEXICOSTATISTICS AS A BASIS
FOR LANGUAGE CLASSIFICATION:
INCREASING THE PROS, REDUCING THE CONS

George Starostin

"Lexicostatistics", a method originally proposed by Morris Swadesh to build relative genetic classifications of languages based on percentages of related items in their basic lexicon, and "glottochronology", used to assign absolute dates of splitting to language groups based on the assumption of a regular rate of change, have not been overtly popular with mainstream comparative linguists, after an early set of critical works had undermined their general credibility. Since then, however, significant process has been achieved in understanding and correcting the flaws of the original method. The current paper focuses on drawing attention to some of these corrections, such as (a) distinguishing between externally and internally triggered lexical change, and (b) factoring out independent semantic innovation. This improved methodology, without significantly cluttering up the formal apparatus, consistently yields results that are not only more credible than Swadesh's original procedure, but are also much more in line with standard comparative-historical linguistics.

INTRODUCTION

As of today, the lexicostatistical method of evaluating degrees of genetic relationship between different languages, based on percentages of historically related items in their basic vocabularies, is already more than sixty years old[1]. Ever since pioneered in the 1950s by Morris Swadesh (Swadesh 1952; 1955), lexicostatistics has had a long and troubled history, rife with criticism, rejection, sometimes even open derision of the method and its supporters. Nevertheless, despite all the problems, controversies, and misunderstandings, we can now state with certainty that the method itself has survived – due partially to the relative ease of its practical application, and partially to its original built-in flexibility, which has allowed researchers to try out different alternate approaches, depending on the scope and nature of the encountered issues.

Furthermore, over the past decade, there has been a veritable explosion of studies on the applicability of network-based models for linguistic classification. Such

---

[1] Certain embryonic forms of the same method can be traced to even earlier times (Hymes 1973), but only Swadesh may be credited for developing and popularizing a fully formalized approach.

studies are more often than not produced by specialists who have had little or no practical experience in the field of historical linguistics, but are nevertheless eager to apply their methods in this area of social sciences as well. From an "evolution-ary" point of view, they are, in and out of themselves, little more than somewhat sophisticated variations on classic lexicostatistics[2]. At the very least, they usually start out with the same idea of standardized wordlists and calculations of cognate percentages between compared languages, to which, of course, they may then apply mathematical algorithms that are significantly different from, and, sometimes, much more complex than the original Swadesh method. Occasionally, these algo-rithms are hard to comprehend, and even harder to put to practical use by historical linguists, many of whom lack the proper computational training to be able to quickly assimilate and evaluate the rapidly growing literature on the subject. Which begs for the obvious question: what exactly *was* so wrong with classic lexicostatistics in the first place, to the extent that it would need to be discarded – and then replaced by far more complicated, and far less practical, alternate models?

Most of the studies that bring out the flaws of the method and suggest correc-tions or alternate models usually approach the matter from a strictly mathematical point of view: "if one formula fails, let us try another". The result is a staggering amount of different models and algorithms, few of which have any noticeable ad-vantages over others once they are applied to a wide range of data, rather than one or two language families that they are usually tested upon. At the same time, these models, with alarming regularity, fail to take into account the enormous theoretical, methodological, and empirical base that has been accumulated in historical linguis-tics over the past two centuries. Statistical methods may have been unjustly ne-glected by the main bulk of historical linguists, yet such methods should not be applied blindly – without a proper understanding of the nature of language evolu-tion, the characteristic differences between vertical (genetic) and horizontal (areal) types of change, and the various mechanisms that stimulate such change on differ-ent levels (phonetical, grammatical, and lexical).

In the light of this statement, the current chapter will focus not so much on the appropriate mathematical apparatus behind lexicostatistics as on the (no less impor-tant) methodological issue – how should one treat the lexical data submitted to lexicostatistical analysis, and how should the results of this analysis influence our historical judgement? Essentially, the main result of every lexicostatistical analysis

---

2   A detailed overview of either the published literature on Swadesh-type lexicostatistics, or the more recent publications on issues of statistical analysis of lexical data for historical purposes, would require a separate chapter all by itself. I will limit myself by simply listing the most obvious, easily available, and comprehensive sources. Embleton (1986) offers a good overview of the history of lexicostatistics up to that point; the extensive 2-volume collection (Renfrew et al. 2000) contains numerous perspectives of leading specialists in the field, both supportive and critical of the method. A thorough comparison of the advantages of tree-based and network-based classification models, fueled by lexical and other types of data, may be found in (McMa-hon & McMahon 2005), along with numerous references to preceding works on the subject. Finally, one of the latest all-encompassing treatments of the same issue is the collective mono-graph (Renfrew & Forster 2006), several of the papers in which attempt to apply statistical and probabilistic methods, carried over from social and natural sciences, to linguistic data.

is a phylogenetic tree (or network). However, different procedures yield different trees, depending on selected types of data, methods of calculation, and the amount of historical information (such as knowledge of regular phonetic correspondences) fed into the algorithm. It is absolutely vital, if these results are to be taken seriously, that the "optimal tree" (or network) be not only selected by means of a formal algorithm, but also tested against historical and typological evidence gathered by linguists – only in this case is there any hope of finally reaching a solid balance between traditional comparative linguistics and modern phylogenetic methods, and, consequently, some hope that our "optimal" tree might have something to do with historic reality, instead of just looking cool on paper or on the screen.

A more or less serious overview of this issue needs to consist of several points. First, we should recall the major theoretical presumptions and methodological implications of classic lexicostatistics and glottochronology. Next, a brief mention is necessary of the most important criticisms that, in the eyes of some specialists, have allegedly "discredited" the methodology beyond repair. I will then try to show how the perceived flaws of the method have been corrected (sometimes, unfortunately, not accompanied by the proper publicity) or may be corrected in the future. Finally and, most importantly, I will attempt to demonstrate that there exists a mutual dependence between general historical linguistics and lexicostatistics, which, if properly recognized and accounted for, may yield phylogenetic results that will be equally pleasing for the statistician and the comparative linguist alike.

THE ORIGINAL METHOD

A general, very concisely stated, interpretation of the major theoretical assumptions of lexicostatistics according to the Morris Swadesh model may be found in (Arapov and Hertz 1974: 21) (an extremely interesting and useful monograph on mathematical methods in historical linguistics, unfortunately, only available in Russian)[3]. It consists of four points, which may be packed into three for the sake of brevity:

[1] Within the lexicon of any language there exists a particular section that may be called "basic" or "stable", so that it is possible to provide a list of meanings which in any language of the world will be represented by words from this section (the so-called "Swadesh list", consisting of 200 items in its large version and of 100 items in its "compressed" version, represents an approximate, somewhat idealized version of this part of the lexicon).

[2] The percentage of words from the basic lexicon which is not replaced by other words over a given time interval is *constant*; it depends only on the amount of time elapsed, and not on any other factors.

---

3   In particular, one of the main ideas of the publication is the existence of a direct correlation between the frequency of usage of a given word and its stability in the language over time, which the authors demonstrate by running a series of tests on select data from Indo-European languages – presaging the widely publicized paper (Pagel et al. 2007) by more than three decades.

[3] All of the words on the Swadesh list are (more or less) equally likely to be re-
    tained or replaced during any particular period of time.

The principal implication is that, by calculating percentages of "cognate" words (i.
e. those known or assumed to have evolved from a common ancestor) on Swadesh
lists of related languages, one may not only arrive at a sound *relative* classification
of these languages in the form of a genealogical tree, but also, having empirically
calculated the rate of lexical replacement on historically verifiable cases, attach an
absolute dating to each of the "splits" indicated on our tree. The assumption of a
constant rate of replacement, therefore, transforms basic *lexicostatistics* (a method
that establishes the relative degree of different languages' proximity to each other)
into what Swadesh called *glottochronology* – a method that builds a linguistic time-
line for the transformation of one protolanguage into an entire group of its present-
day or historically attested descendants, based exclusively on data provided by the
languages themselves.

Since the original method was essentially inspired by and founded on the same
basic principle as radiocarbon dating, it is hardly surprising that it may be encap-
suled in the same formula: $N(t) = N_0 e^{-\lambda t}$, where $N(t)$ is the share of original words
left on the wordlist after an elapsed period of time t (from 1 to 0), N_0 is the size of
the original list, and λ is the replacement rate. For the 100-wordlist, the rate was
originally calibrated by Swadesh as ≈ 0.14, i.e. 14 replacements out of 100 per mil-
lennium.

Apart from a vague appeal to the necessity of preserving mutual intelligibility
between different generations, Swadesh himself never offered a theoretical rationale for the notion of a regular rate of change for the lexicon. However, there seemed
to be plenty of empirical evidence for this assumption, accumulated from lexical
comparison of modern day related languages (for instance, almost any two languages taken from different branches of Indo-European seem to always yield
around 25 to 35% of cognates on the 100-wordlist), as well as calibrations of the
method on those few languages whose history was known over a period of two to
three thousand years (most of them also belonging to the Indo-European family).

CRITICAL REACTION

Despite some initial interest and acceptance on the part of some linguists, Swadesh's
glottochronology quickly acquired an overall negative reputation that eventually
almost succeeded in making it into a bad word in the linguistic community (to the
extent that even some of the newer approaches to inferring separation dates from
linguistic evidence, such as the one advocated in (Gray and Atkinson 2003) and
further developed in subsequent publications, fell under the same wave of criticism,
forcing their authors to explain, over and over again, that "this is *not* glottochronology!" – even if there is really nothing wrong with expanding the use of the term to
denote any procedure that deals with linguistic dating).

It should be mentioned that Swadesh himself never drew a firm line between
"lexicostatistics" and "glottochronology", and neither did many of his critics (thus,

e. g., in (Campbell and Poser 2008: 167): "the term 'lexicostatistics', while given a technical distinction by some, is usually used as a synonym of glottochronology"). Indeed, it would seem reasonable that phylogenetic classifications based on lexicostatistics only make proper *historical* sense under the condition that the idea of certain *rates* of lexical change is involved – be it a permanently constant rate, or one that could be averaged through rate-smoothing algorithms, or one depending on particular external factors and parameters that can be integrated within the model. That said, it is always possible to perform lexicostatistics as such, without insisting on a chronological interpretation of the resulting trees (or networks), and, from this point of view, "glottochronology" is an extension of "lexicostatistics". All major criticisms of the Swadesh method may consequently be divided into one group pertaining to "purely lexicostatistical" aspects of the procedure, and another group that is directly targeting the chronological aspect.

Detailed overviews of all the critique may be found in previously mentioned sources, most notably (McMahon and McMahon 2005). A compact listing of the most frequently raised issues might look something like this:

1) *Impracticability* of lexicostatistics as such. Almost from the outstart, many specialists in particular language families found it hard, or even impossible, to construct uniform Swadesh lists for their areas of linguistic expertise, because the "semantic concepts" represented by English words on the list were occasionally found to be lacking in the respective languages (e. g. the absence of a word for 'horn' in Polynesian languages, or of a word for 'fish' in many Bushman languages, etc.), or, much more frequently, found to be too vague and broad, allowing for the (sometimes obligatory) choice of multiple synonyms instead of one single word (e. g. the lack of a single term for 'to eat' in many African and American languages that distinguish lexically between 'eating soft food', 'eating meat', 'eating nuts/fruit/etc.'). (Hoijer 1956) is a good early example, illustrating the practical difficulties of reconciling the actual lexical data (of Navajo) with the somewhat arbitrarily established lexicostatistical "standart".

2) *Falseness of assumption* [1]. Based on the gradual progress of areal linguistics (the study of convergent processes in languages), many linguists have pointed out that the "stable" items on the Swadesh list are actually not as "stable" as they might once have seemed; in other words, that lexicostatistics heavily underestimates the role of language contact, placing too much emphasis on "words" as genetic markers. Some even question the necessity of distinguishing between the concepts of "basic" ("stable") and "cultural" ("unstable") lexica as such, e.g. (Haarmann 1990), denying their universal application.

3) *Falseness of assumption* [2]. This criticism concerns the glottochronological application of lexicostatistics. Simply put, it denies the existence of any constant rate of lexical change – not on theoretical grounds, which would be speculative, but based on actual evidence accumulated from several test cases. The "textbook" example of this criticism is the famous paper (Bergsland and Vogt 1962), whose test case of Norwegian (too high) vs. Icelandic (too low) rates of change played a significant part in discrediting glottochronology, but other im-

portant works may be quoted as well, such as Robert Blust's research on the
varying rates of change in Austronesian (Blust 2000).
4) *Falseness of assumption* [3]. Regardless of whether one accepts the idea of a
    constant rate of change or not, it is more or less obvious to anyone acquainted
    with a large enough amount of comparative data that some words on the
    Swadesh wordlist are, on the average, more stable than others. For instance,
    personal pronouns such as 'I' or 'you', all over the world, generally get re-
    placed much less frequently than such words as 'small' or 'yellow'. This im-
    plies the necessity of some sort of *grading* procedure for the wordlist, not orig-
    inally considered by Swadesh. The importance of this issue was, among others,
    stressed by the late Sergei Starostin (1989; 2000).
Another important criticism has been added to the overall stack only recently, due
mainly to the rapid growth of areal linguistics, and partly to the interdisciplinary
influence of other branches of science:
5) *Insufficiency* of the lexicostatistical method to yield a proper model of all the
    historical connections between compared languages, since lexicostatistics can
    only result in a tree-like structure, which (allegedly) does not always corre-
    spond to historical reality. (cf., e.g. Bateman et al. 1990 and many other similar
    works).
Overall, it may be assumed with reasonable safety that the major "irritating factor"
of lexicostatistics/glottochronology over the years has always been the idea of a
constant rate of lexical change; all other criticisms bear a prominent technical nature, and could, at least in theory, be overcome through careful refining and calibra-
tion of the method. Criticism (2), however, upon first impression seems to be striking at its very heart. Indeed, even a single unexplainable exception from the "rule"
drastically undermines its usefulness, placing heavy doubt on all lexicostatistical
classifications, regardless of how realistic or compatible with other types of data
they may look on their own[4].

And yet, it seems that the disagreement over the "rates of change" issue is actually only one facet of a much larger problem, only one consequence of a significant
misunderstanding between various proponents and opponents of the lexicostatistical method. This misunderstanding is perhaps best illustrated by a brief passage in
(Campbell and Poser 2008: 167–8), where "glottochronology" is listed as an *au-
tonomous* "method" of testing language relationship – right next to Joseph Greenberg's "multilateral comparison" as another such "method". The exact same misunderstanding, in a very concise (and somewhat brutal) manner, is ensconced in an
earlier critical paper by Alexander Vovin: "I certainly do not subscribe to the notion
that one always can use 13 or even 100 basic vocabulary items to prove a genetic
relationship[...] but I can well understand that using glottochronology for proving
genetic relationships is really compelling: all you have to do is just to compare 100
words taken from dictionaries. Easily done, and the results are overwhelming" (Vovin 2002: 164).

---

4   Cf.: "The central question with regard to the validity of lexicostatistics has never been "can the
    mathematics be improved to make the method 'work'?" Rather, it has been "is the 'universal
    constant' hypothesis empirically justified?" (Blust 2000: 326).

Essentially, in these and other works critical of the method "lexicostatistics" is often viewed as something that has no obligatory ties whatsoever to the classic comparative method in historical linguistics. Sometimes this idea is presented as the "current" (corrupt!) state of lexicostatistics, as opposed to the "original" Swadesh incarnation of the method, in which it was designed to be used exclusively as a means of establishing the relative degrees of genetic relationship between languages that have already been proven to be related by more conventional means; i.e. "genuine" lexicostatistics cannot *create* a genealogical tree, it can only serve as one way of establishing the correct configuration of its nodes and measuring their length. But in either case, much of the mistrust that comparative linguists feel towards lexicostatistics stems from a general mistrust in doing historical linguistics "by the numbers": driven by the old slogan that "each word has its own history", specialists automatically view any reductionist attempt to fit the immense world of lexical change into a simple formula with suspicion, and are only too happy to embark on a search for rule-breaking exceptions, regardless of whether these exceptions in themselves may be explained by further conditioning.

In order to sort out this confusion, I believe that, first of all, it is important to stress, as concisely and transparently as possible, the following three points:

1. A sharp distinction must be drawn between **preliminary lexicostatistics**, in which cognacy judgements are made based on *phonetic similarity* of the compared items, and **proper lexicostatistics**, in which cognacy judgements are made based on *regular phonetic correspondences*, established between compared items (based on additional morphemic data as well, not solely those words that belong to the Swadesh wordlist).

The methods, goals, and results of these two procedures are significantly different from each other. Preliminary lexicostatistics is a useful, although not highly conclusive, procedure during the initial stages of historical research on a potential language relationship. Its major advantage is that it can be based on completely objective standards or even rendered completely automatic. For example, the StarLing software, developed by Sergei Starostin, today includes a handy plugin that analyzes the phonetic structures (more precisely, "consonantal skeletons") of all the words on processed wordlists, then assigns "pseudo-cognacy" indexes to words whose basic structures are similar enough to be considered matching. Thus, the word *mata* in language A would be considered "cognate" with the word *meda* in language B, since both can be reduced to the basic consonantal skeleton MT, but not with the word *maka* in language C, whose skeleton has the structure MK (see Starostin (2008) for more details). Another example of the "automated" approach is the ASJP (Automatic Similarity Judgement Program) project, run by several specialists at the Max Planck Institute; its basic algorithm is slightly more complicated, using the Levenshtein distance method to arrive at "pseudo-cognacy" judgements, but the results are not necessarily more reliable than the ones produced by the StarLing plugin.   ·

The goals of preliminary lexicostatistics, however, are neither to "prove" language relationship as such, nor to acquire reliable chronological information on the history of the putative "family". Preliminary lexicostatistics can be performed on any languages, no matter how distantly related – English and Kiswahili will work as fine as any other pair – and serves, accordingly, to formulate *preliminary* hypotheses on relationship and classification, which can thereupon serve as research fuel for the comparative linguist. To carry out this procedure, it is indeed sufficient to "compare 100-wordlists taken from dictionaries"; but no comparison of any number of 100-wordlists *per se* can lead to definitive conclusions (unless the compared languages are extremely close to each other, e.g. on the level of Slavic or Turkic; but there is actually no point in subjecting such language groups to preliminary lexicostatistics in the first place)[5].

2. "Proper" lexicostatistics is the only application of the method that can aspire to conclusive results, and it is not *opposed* to the comparative method, but should be viewed as a *complementary* technique. Ideally, the reconstruction of a protolanguage on the basis of the classic comparative method should always be accompanied with a lexicostatistical check, since there is a mutual benefit between the two. Lexicostatistics that is not based on the findings of the comparative method is not "proper" (and, therefore, inconclusive), whereas the comparative method without lexicostatistical support lacks a proper quantitative foundation.

The crucial importance of lexicostatistics emerges with particular clarity when it comes to objective assessment of questionable relationship hypotheses that, on the surface, claim to be based on the comparative method, but in reality explore its "holes" (such as, for instance, the lack of precise standards for semantic reconstruction) to produce unrealistic protolanguage systems. A routine lexicostatistical check that verifies to what extent the phonetic correspondences, proposed for these systems, are actually applicable to the basic lexicon of the compared languages, can quickly and quite convincingly weed out most of the false (or, at least, "undemonstrable") hypotheses. (For just a few examples, see Starostin (2002) on such a demonstration for J. McAlpin's "Elamo-Dravidian", or Kassian (2010) on a similar demonstration for A. Bomhard and A. Fournet's "Hurro-Indo-European").

3. It is absolutely imperative that the essence of lexicostatistics should not be reduced to discussions on its mathematical representation. Unlike genetics, where researchers operate with huge numbers of characters that can only be properly assessed within the framework of general models, Swadesh-type wordlists are generally small (100–200 items), stimulating individual case studies of the evolution of particular meanings in particular languages. "Proper" lexicostatistics derives from etymological judgements made by historical linguists, but etymological judgements are often questionable. Some of them may be completely false, being based on erroneous phonetic correspondences; some may be mistaking areal contacts for cognacies; in quite a few cases, "cognacies" may be etymologically correct, but reflect

---

5   Unfortunately, Swadesh himself contributed to this confusion. His early works, which introduced lexicostatistics and glottochronology to the general public, mostly dealt with the "proper" method, operating on language groups and families that had already been firmly established (such as Indo-European) to determine the internal classification of these units. In later works, however, he would occasionally demonstrate a subtle transition to "preliminary" lexicostatistics, using the method to justify a belief in "Dene-Finnish" and similar far-flung relationship hypotheses, e.g. in (Swadesh 1965).

the result of unilateral independent semantic developments (see below) rather than direct descent of the "form/meaning" pair from the nearest common ancestor. All of these problems influence the outcome of the calculations and sometimes result in significant errors.

In the light of this, I am personally less interested in whichever of the various formulae/algorithms suggested, at one time or other, to improve the quantitative basis for lexicostatistics, is "better" or "worse", than in the actual mechanisms, laws, and tendencies of linguistic change that this method attempts to uncover, and in the actual general problems of comparative linguistics that it pulls out into the limelight from under the rug where they have been, way too often, conveniently swept by "traditionalists".

Three of these problems may be considered critical for any further application of lexicostatistics (and, in fact, for the future development of historical linguistics on the whole). These are: (a) the issue of *synonymy* in lexicostatistical calculations and semantic reconstruction; (b) the necessity of strict differentiation between *internally* and *externally driven lexical change*; (c) the phenomenon of *unilateral independent semantic development* and its particular importance for deep level reconstruction.

The first of these issues has been recently discussed at length in (Starostin 2010), where a much stricter approach to the selection of synonyms than is usually being adopted in lexicostatistical studies has been advocated for; a set of possible practical "guidelines" for such selection has also been published as (Kassian et al. 2010). However, within the scope of the current volume this problem is not as relevant as the other two, which bear a direct connection to the much larger issue of distinguishing between shared "horizontal" (areal) and "vertical" (genetic) features, as well as the question of tree-based vs. network-based models in historical linguistics and their relative flaws and advantages. For this reason, the remaining parts of the chapter will concentrate on the "theoretical underbelly" of these two issues, and on the practical implications that they carry for the active lexicostatistician.

INTERNALLY DRIVEN VS. EXTERNALLY DRIVEN LEXICAL CHANGE

Much of the traditional animosity towards the glottochronological application of lexicostatistics had, and still has to do with the alleged "constant" or "regular" character of the rate of lexical change. Specialists have often rejected the idea on theoretical grounds, claiming that no such "regularity" is at all possible when we are dealing with such an unpredictable object as language, changing at the whim of whatever social and historical factors it may encounter, e.g. (Haarmann 1990). However, the criticism is particularly biting when the critic in question is armed with empiric arguments, e.g. Bergsland & Vogt with their evidence for different rates of change in Norwegian and Icelandic (or R. Blust with evidence from Austronesian).

It has already been noticed, many times, that "rates of lexical change" are particularly fluctuating under specific conditions – namely, the presence of a strong

outside linguistic influence, resulting in numerous borrowings from one language
to another over, sometimes, a very short time interval. In particular, the issue was
tackled by Sergei Starostin (Starostin 1989), who has, in his analysis of the "Bergsland & Vogt controversy", demonstrated that Icelandic and Norwegian actually
show comparable rates of change once the numerous borrowings into literary Norwegian from other Germanic languages have been excluded from lexicostatistical
calculations. Likewise, it is quite evident that borrowings have also drastically sped
up the rate of lexical change in such languages as Albanian, Brahui (Dravidian family), Northern Songhay (a seriously "Berberized" variety of Songhay), and there is
a significant probability that the different rates of change, observed for different
Austronesian languages, depend first and foremost on the degree of contact between these languages and the neighboring non-Austronesian idioms (Peiros 2000).

These observations imply that it is absolutely necessary to distinguish between
two types of lexical change, which may be respectively called *internally driven
change* (IDC) and *externally driven change* (EDC). The first of these occurs when
words are replaced "from within" the language, i. e. from the inherited lexical stock.
Specific factors that trigger such replacements are generally obscure (and probably
quite numerous). However, the more we study particular cases, the more evident it
becomes that IDC is almost never motivated – more precisely, cannot be demonstrated to have ever been motivated – by specific social factors (with the probable
exception of occasional taboo usage). Instead, the general mechanism is rather one
that could be called the "**aging of words**": essentially, the further a word persists in
the language, the higher are its chances of being replaced by a different word in the
next generation of speakers.

The reasons behind such "aging" are anything but mystical: it reflects the process of *polysemization*, i.e. the acquiring of figurative or metonymically adjacent
meanings by the original word[6]. Universally typical examples, independent of cultural specifics, include such developments as 'head' → 'top', 'beginning', 'origin',
'chief', etc.; 'hand' → 'handle', 'protruding part', 'help', 'assistance', etc.; 'tree'
→ 'wood'; 'water' → 'flowing water', 'river'; 'fire' → 'heat', 'sun', 'bonfire', etc.
The more polysemous the word becomes, the higher is the pressure on locating se-
mantically similar words that could start to express its *original* meaning: the tendency to acquire additional meanings becomes counterbalanced by the opposite
tendency to decrease the resulting linguistic ambiguity that hinders successful communication.

Most importantly, the more we study semantic typology and the various types
of meaning shifts around the globe (at least, on the experimental level of the
Swadesh wordlist), the more it becomes clear that, on the average, these processes
happen in similar, almost *uniform* ways all over the place. This, in turn, could imply
that it would be reasonable to expect the average rates of the accumulation of such
changes to be comparable as well.

---

6   Certain embryonic forms of the same method can be traced to even earlier times (Hymes 1973),
     but only Swadesh may be credited for developing and popularizing a fully formalized ap-
     proach.

Externally driven change, on the other hand, is completely unpredictable. Although some of the items on the Swadesh list are known to be less prone to borrowing than others (see Haspelmath and Tadmor (2009) for the so-called "Leipzig-Jakarta" wordlist, alternative to Swadesh, that attempts to take this factor into consideration), this tendency does not work too well in situations where one language has become subject to "massive lexical bombardment" on the part of another. Such languages as Brahui (≈ 30% of borrowings on the 100-wordlist from several surrounding languages), Albanian (≈ 25% of borrowings), Tadaksahak Songhay (also ≈ 25% of borrowings from Tuareg), etc., give the impression of being capable of borrowing from any type of lexical layer almost at random.

Two solutions have been offered so far in dealing with this problem. One was to attempt to "calibrate" the results by introducing the idea of "borrowing rates" (e. g., in Embleton (1986), where several previous attempts, mainly by A. Dobson and D. Sankoff, are also mentioned). However, such calibrations only really make sense in situations where borrowing occurs on a gradual basis – two or more languages slowly diffusing lexical items among each other – whereas most of the major issues with lexicostatistics and glottochronology concern "explosive" situations that exclude the very idea of a "rate": in the three examples quoted above, for instance, there is every reason to believe that the majority of the borrowings took place over a relatively brief period, certainly not exceeding 1,000 years.

The other solution, supported in S. Starostin's model, was to altogether exclude borrowings from lexicostatistical calculations; in other words, the regular rate of change has been limited to include *only* IDC. This, of course, presumes that the lexicostatistician always knows what exactly has been borrowed and what has been inherited – which is, in most cases, impossible. For instance, while we are quite certain that the English word *mountain* has relatively recently been borrowed from French, the origins of the word *dog* are far more obscure: tentative derivations from a Proto-Germanic source are not very convincing, leading to possible speculations about borrowing from a pre-Anglo-Saxon substrate – direct evidence for which is, however, lacking. The case is much worse for poorly studied languages all over the world that do not have any written history at all.

It must, however, be kept in mind that the division line between IDC and EDC, like most division lines in linguistics, is sometimes rather blurry. For instance, a word may be borrowed from language A into language B with one meaning that will later gradually evolve into another – which, coincidentally, will be a "Swadesh meaning". Such is the case with the well-known Romance word for 'liver' (Italian *fegato*, French *foie*, etc.), going back to Vulgar Latin **ficatu* 'fig-stuffed' (liver). The Latin word for 'fig' itself was borrowed from a Semitic source; yet, by the time it began replacing the original word for 'liver' (*iecur*) in its descendants, it was a fully assimilated word whose semantic shift was clearly a case of IDC rather than EDC.

On the surface, it might seem that such difficulties make the task of separating IDC from EDC technically impossible in way too many cases. But in actuality, in order for the lexicostatistical procedure to work properly, there is no need to achieve a complete and certified separation. All that *really* needs to be set up is a filter to weed out *massive* EDC – of the type represented by Norwegian, Albanian, Brahui, Northern Songhay, etc. Results for languages whose basic lexica are generally "loanproof" and only sporadically shift due to external influence (in other words, the absolute majority of the world's languages) will not differ drastically, regardless of whether we have or have not excluded all the identified loanwords from our calculations.

Thus, English and German, once we exclude loanwords from Romance and other Germanic languages from calculations, share approximately 82% cognates on the remaining part of the wordlist. With loanwords included, this number drops down to approximately 78%, yielding a glottochronological date of separation that corresponds to 250 AD rather than the more "correct" 500 AD; still, the difference is minimal and lies well within acceptable margins of error. On the opposite side, however, if we try to do the same thing with such Dravidian languages as Tamil (whose basic lexicon shows a modest, but significant number of borrowings from Sanskrit) and Brahui (which has borrowed up to 30% of its basic lexicon from Baluchi, Hindi, Arabic, and Persian), the number drops down from about 38% to approximately 26%, yielding dates of separation that roughly correspond to 3000 BC and 2000 BC, respectively – with various types of comparative evidence indicating that only the first one of these may be close to the truth.

It becomes, therefore, the primary duty of the lexicostatistician to be able to locate and diagnose those languages that, due to particular sociolinguistical factors, are capable of wrecking havoc on their basic lexicon within a short period of time (an appropriate term for this phenomenon would be something like "*low lexical immunity*"). This task, although extolled in quite a few critical works as extremely problematic, mainly depends on the quality of descriptive materials available for any given linguistic area.

As for internally driven change, for the reasons listed above, I see no significant theoretical reasons that would prevent it from showing a generally regular speed pattern throughout the centuries. At this point, it still remains to be seen that rates of IDC may significantly vary from each other. In fact, I would say that lexicostatistics and glottochronology are reasonably free from the threat of total extinction as long as the following situation has not been explicitly demonstrated on uncontroversial data:

– comparison of unquestionably related (relationship proven based on phonetic correspondences, morphological evidence, etc.) languages A and B shows that A has undergone n internally driven replacements, while B has undergone $2 \times n$ internally driven replacements over the same time interval, where n is no less than 4 or 5 (to rule out statistical margins of error).

One or more such demonstrations would surely put an end to all debate about rates of lexical change; however, I know of no such examples, despite having closely worked with close to a thousand different Swadesh lists from various families, and have a strong suspicion that none will be discovered in the near future. Until such a discovery is made, the "myth" of regular lexical change will continue to lay a strong claim to reality, and lexicostatistical models based on this empirical assumption will continue to be useful.

UNILATERAL SEMANTIC DEVELOPMENT
AND ITS PRACTICAL IMPLICATIONS

In the remaining part of this chapter, I would like to draw attention to a phenomenon that seems to be rarely, if ever, discussed in theoretical works on comparative linguistics, despite its significant importance, especially in the area of the so-called "long-range" comparison that deals with taxonomic units of considerable time depth ("macrofamilies").

One of the main reasons for replacing tree-based models in historical linguistics with network-based models, advocated by a steadily growing number of specialists, is the "loss of information" argument: admittedly, tree-type classifications do not fully reflect all the types of historical relations between languages, since they can only reflect language divergence, with "vertical" transmission of linguistic heritage, but tell us nothing about the elements of convergence and "horizontal" transmission. From a purely theoretical point, there is nothing new about this approach: the importance of recognizing convergence, dialectal mixture, and areal influences as important factors in language development, and the necessity to come up with models that formally reflect this recognition was clearly stated already in the XIXth century (e. g., by J. Schmidt and his "wave theory"). What *is* new is that the network theory has managed to properly formalize this theory, advancing from the highly approximate and impractical "wave diagrams" of Schmidt to complex objective representations based on uniform datasets (such as the Swadesh wordlist, or computerized etymological databases).

Networks are, however, generally harder to interpret than trees — especially if we share the belief that neither trees nor networks in historical linguistics should be a goal in itself (i. e. work as a "pretty picture" that looks nice as part of a Power-Point presentation), but must rather work as useful approximations, suggesting an optimal historic scenario for the gradual transformation of one protolanguage into a set of descendants. The big advantage of a tree is that each tree has a unique historical interpretation, whereas each network conceals a variety of scenarios. The disadvantage is, of course, that each particular tree may not only be "incomplete" as a reflection of language history, but may even be "wrong" (for instance, mistakenly grouping distantly related languages together as close relatives due to erroneous cognacy judgements or undetected borrowings), whereas it is not clear if a network, whose very purpose is to suggest a variety of alternative scenarios, may ever be "wrong" as such.

In fact, networks seem almost to be an unavoidable necessity. Trees are constructed from studying the fate of "characters" (= words on the Swadesh list or other sets of data), which can frequently be different enough to allow for various schemes of branching. Consider, for instance, the equivalents of two different Swadesh meanings in three Indo-European languages (indexes A and B indicate formal cognacy, i. e. whether these lexical stems go back to a common Proto-Indo-European ancestor or not):

	Meaning	Hindi	Irish	Tocharian A
(1)	'to kill'	mār-$_A$	maraim$_A$	ko$_B$
(2)	'ear'	kān$_C$	cluas$_D$	klots$_D$

Converted into tree form, character (1) could suggest a branching into Tocharian vs. "Hindi-Irish"; character (2) would, however, suggest the exact opposite – a branching into Hindi vs. "Irish-Tocharian". From a purely formal standpoint, we would have little choice but to superimpose these two trees onto each other, getting a network representation, similar to the usual way this is done in genetics. But such a projection would leave us no closer to answering the most important question: which of the two choices, A/C or B/D, were actually used to express these meanings in the common ancestor of all the three languages? Somebody with no knowledge whatsoever of Indo-European historical studies would list the following possibilities:

(a) *both* A and B could mean 'kill' and 'ear' in the proto-language, with each daughter language retaining only one synonym out of the two. This is highly unlikely, since it goes against the uniformitarian principle: normally, each language is supposed to only use one word in any given Swadesh meaning, and there is no reason to surmise a different picture for reconstructed protolanguages[7]. In general historical linguistics, such accumulations of synonyms in proto-languages can only be regarded as "cop-outs" substituting for genuine semantic reconstruction;

(b) the most widely distributed roots (A and D) had the meanings 'kill' and 'ear' in the proto-language, with Irish remaining as the most "conservative" descendant and the other two languages each sharing one innovation. Since only shared innovations are diagnostic of branching, the resulting tree would have all three languages as equidistant;

(c) the *least* widely distributed roots (B and C) had the meanings 'kill' and 'ear' in the proto-language. If this were true, we would have one shared innovation (A) between Hindi and Irish and one more (D) between Irish and Tocharian – an extremely "un-tree-like" situation that a traditional comparative linguist would tend to avoid;

(d) a mixture of (b) and (c) – for instance, Hindi-Irish 'to kill' could be a shared innovation, reflecting a binary branching into Tocharian vs. Hindi-Irish, whereas Irish-Tocharian 'ear' could be an archaism, opposed to a lonesome innovation in Hindi, reflecting nothing in particular.

Let us now look at the larger picture. Although the general cognacy percentages between all three languages are very similar (around 25% for each pair), to the best

---

7   This is actually a well-known debatable point, but there is not enough space in the chapter to present the full argumentation in its favor. Suffice it to say that, if the "Swadesh meaning" is understood as a very narrowly defined basic notion, restricted to certain syntactic contexts and not including additional semantic or stylistic components, the task of correlating it with one and only one equivalent in any given language becomes much easier than is sometimes complained about in critical literature; see (Starostin 2010) for more details.

of my knowledge, no tree structures have been proposed for Indo-European on which Irish (and other Celtic languages) would come out closer to Tocharian than Hindi (and other Indo-Aryan languages). On the contrary, most lexicostatistical (and not only lexicostatistical) classifications usually list Tocharian as one of the earliest branches to "lop off" from the common Indo-European stem. This would imply that the common ancestry of Irish *cluas* and Tocharian *klots* is, most likely, an archaism: since the overall mass of the evidence is against a "Celtic-Tocharian" branch, their sharing a lexical item against Hindi (and other Indo-Aryan languages) can only be reasonably explained as preservation of a bit of Proto-Indo-European heritage, irrelevant for classificatory purposes.

But here is the catch: careful etymological analysis of the evidence shows us that Irish *cluas* and Tocharian *klots* do *not* preserve the original Indo-European word for 'ear'. That word is almost certainly represented by an entirely different root, the one found today in such forms as Russian *yx-o*, Lithuanian *aus-is*, English *ear*, French *or-eille* (← Latin *aur-iculum*), etc., going back to Proto-Indo-European *ous-*. Unlike the Irish and Tocharian forms, this root is found in a much larger number of branches, and, most importantly, it is unmotivated, i.e. represents an original non-derived nominal stem, whereas both *cluas* and *klots* may uncontroversially be regarded as nominal derivatives from Proto-Indo-European *kleu̯-* 'to hear'.

This puts us in a difficult situation close to the one suggested in (c): if 'ear' in Proto-Indo-European was most likely *ous-*, and Irish and Tocharian do not form a single node on the tree, how do we explain this shared innovation between them? One way out would be to suggest a trace of contact – a "hidden borrowing", perhaps, from one language branch to another. No historical linguist, however, would take seriously the possibility of contacts between Irish and Tocharian: at best, one could think of such a possibility for some very early stages of Proto-Celtic and Proto-Tocharian, during which they may not yet have been separated by thousands of miles, but even then the situation would border on the comical – why in the world would these two groups of speakers of early Indo European dialects want to influence each other in their choice of the basic equivalent for 'ear' and nothing else? (Exclusive Celtic-Tocharian lexical and semantic isoglosses are quite rare, to say the least).

There is only one other solution: assume that the shift from *ous-* to *kleu̯-* took place *independently* of each other in Irish and Tocharian. How high is the probability of that assumption? If we remember what has already been mentioned above on the issue of semantic change typology – namely, that some types of meaning shifts happen far more frequently than others – it must be quite high, since the semantic shift from 'hear' to 'ear' is one of the most commonly encountered shifts connected with these meanings all over the world (curiously, the *opposite* shift, from 'ear' to 'hear', is extremely rare in comparison). Roughly speaking, if the word *ous-* 'ear' were to be replaced at all in various Indo-European languages, there is no way that there would not have been at least one or two of them in which it were to be replaced by a descendant of Indo-European *kleu̯-* – and, in this particular case, these two languages happened to be Tocharian and Irish. (In fact, Old Irish still has *au* – clearly confirming the hypothesis).

The importance of this process, which we may call *unilateral independent semantic development* (UISD for short), should not be underestimated. For some reason, it seems to be ignored in most works on lexicostatistics (or, at least, is never paid all the attention that it deserves). However, examination of the Swadesh wordlist for Indo-European languages alone shows that UISD may be reliably postulated for many more cases – coincidentally, the Hindi-Irish isogloss for the word 'to kill' (Hindi *mār-* : Irish *maraim*), listed above, also happens to be one such case, reflecting an old Indo-European causative form of the verb **mer-* 'to die', which was certainly not the default equivalent for the meaning 'to kill' in Old Indian.

Admittance of the existence of UISD puts the lexicostatistician in a difficult position, because it introduces an element of ambiguity into the very notion of "cognac". Normally, two words are required to be marked as "cognates" if they go back to the same common ancestor both in form *and* meaning. Russian *yxo* and English *ear* are crystal-clear "cognates", because their phonemic structures exhibit regular correspondences, their basic meanings coincide, and they may be shown by the comparative linguist to directly reflect Proto-Indo-European **ous-* that also had the exact same basic meaning. But what about *cluas* and *klautso*? Their phonemic structures also coincide (at least, as far as the root is concerned), their meanings are identical, but the Indo-European word that they go back to must have, by all accounts, had a *different* meaning. Neither of them continues a "form/meaning" pair that goes back to the same common ancestor; in all likelihood, they represent the results of randomly coinciding paths of development, and, what is most important, we have at our disposition a real instrument to show that this is the most likely situation – distributional analysis of the various forms for 'ear' in Indo-European languages, to which we may add knowledge of the typology of semantic change (a common Irish-Tocharian isogloss reliably deriving 'ear' from, e.g., 'cockle-shell' would be far more difficult to interpret as UISD than the typologically common derivation of 'ear' from 'hear').

Now that we have analyzed the situation, it is clear that cases like *cluas* and *klautso* certainly cannot be "cognates" in the same sense as *yxo* and *ear*. (We may distinguish between "etymological cognacy" of the items, whose forms go back to one and the same protoform, and "lexicostatistical cognacy", when their meanings go back to the same meaning in the protolanguage as well). It may then be useful to mark them with different "cognation indexes" in the database, so that the percentage of true cognates between Irish and Tocharian be closer to the truth. But it is also clear that this correction cannot occur during the "main" stage of proper lexicostatistics. Before understanding that *yxo* and *ear* are "lexicostatistical cognates", whereas *cluas* and *klautso* are not, we need to already have an established tree structure, one that assigns Irish and Tocharian to different branches. And that structure, in turn, is itself created on the basis of a Swadesh wordlist where all the cognacies have already been marked.

Instead of regarding the situation as a sort of vicious circle, I prefer to view it as a variety of bootstrapping, where lexicostatistical analysis alternates, over and

over again, with standard comparative research. From "preliminary" lexicostatistics, helping us to accumulate "raw" comparative data, we advance to the stage of
etymological research, establishing regular phonetic correspondences. Lexicostatistical analysis is then repeated, this time in a "proper" manner, which allows us to
come up with a generally reliable classification scheme. The comparative evidence
is then checked once again for identifiable cases of UISD. Finally, where such cases
have been identified, "etymological cognac" on the Swadesh lists (of the *cluas/
klautso* type) is eliminated, in order to produce a (presumably) even more precise
variant of the tree.

UISD AND LEXICOSTATISTICS IN "LONG-RANGE" COMPARISON

It is possible that the seriousness of UISD as a disturbing factor in lexicostatistical
classification might have been overlooked, were it not for the gradual advances in
semantic typology and the growing awareness of the idea that "meaning shifts" are
not nearly as unpredictable and tremendously numerous as the existing literature on
the subject would have one believe. If any one given meaning at any one given time
may evolve into 1000 different adjacent meanings, the probability of UISD any-
where, at any time, is quite low. But the real situation is different: the real probability of the meaning 'eye' to develop out of the meaning 'see', judging by the accumulated evidence, is much higher than its probability to develop out of *any* other
meaning. If we are looking for semantic parallels to the word 'tongue' in other
languages, the first place to look would be the verb 'to lick'. The meaning 'red' is
more often derived from 'blood' than from anything else. The meaning 'tree'
(growing), with time, is highly probable to develop the polysemy 'tree/wood', and
then lose its original meaning, etc. etc.

Where the scope of this problem becomes really frightening is in "long-range"
comparison, i.e. attempts to establish genetic relationship on deep chronological
levels (exceeding 6000–7000 years), when they are propped up with lexicostatistical support. As an example, one could quote (Starostin 2003), a paper that tries to
verify several long-range hypotheses for language families of Eurasia based on
lexicostatistics. The comparison operates on proto-roots, reconstructed for Indo-
European, Uralic, Kartvelian, Altaic, Dravidian, Semitic, North Caucasian, Sino-
Tibetan, and Yeniseian protolanguages (with varying degrees of reliability). The
comparison itself fluctuates between "preliminary" and "proper" lexicostatistics,
since regular systems of correspondences have been offered for some combinations
of these families (e. g. the "Nostratic" etymological hypothesis of V.M. Illich-Svi-
tych, uniting the first six of the listed families, or S. Starostin's own "Sino-Caucasian" that joins together the last three), but not for others (proper regular correspondences between "Nostratic" and "Sino-Caucasian" have not been established,
and any hypothesis of a high-level connection between these nodes can only be
extremely vague at the present).

Upon first glance, the number of "cognates" accumulated between the languages is quite impressive: the resulting lexicostatistical matrix yields 26% be-

tween Indo-European and Uralic, 35% between Indo-European and Altaic, 28% between Altaic and Dravidian, a staggering 54% between North Caucasian and Sino-Tibetan, etc. Although some of the individual etymologies are questionable on phonetic grounds, such numbers would seem to clearly support not only the very fact of relationship between these families, but even a rather surprising *closeness* of this relationship: for comparison, 30% is the expected average of lexicostatistical matches between modern Indo-Aryan and Iranian languages.

A closer look, however, reveals that a typical comparison between the reconstructed protolanguages included in S. Starostin's tables looks like this:

	Indo-European	Uralic	Kartvelian	Altaic	Dravidian
skin		kopa		k`āp`a	
skin	twak-		ṭqew-		tok-

or like this:

	Indo-European	Uralic	Kartvelian	Altaic	Dravidian
root	wərəd-				vēr-
root		sär-	ʒir-		sīr-
root		ontV		ŋiūnt`e	

The first of these entries groups Uralic together with Altaic against Indo-European + Kartvelian + Dravidian; the second, however, roughly contradicts this scheme, offering three overlapping isoglosses that are neither interpretable in a "tree-like" structure nor offer a clear hint at whichever of these three "cognate groups" could actually correspond to the main lexical equivalent for 'root' in the common ancestor of these language families (Proto-Nostratic).

Opponents of long-range comparison would probably interpret these contradictions as confirmation of the fallacy of Nostratic and similar hypotheses: since we do not usually get that much overlapping, or that many "proto-language synonyms" for well-established low-level families such as Indo-European or Uralic, this picture merely reflects the results of chance. S. Starostin's approach to synonimity (like the one we see in between Uralic *sär-* and *ontV*, both of which represent the meaning 'root') is formulated as follows: "A word can be used as representing a particular meaning in the protolanguage if it has exactly this meaning in at least one subbranch of the family" (Starostin 2007b: 807). This is risky, since a family can consist of quite a few subbranches; if our comparison is not really between Proto-Indo-European and Proto-Uralic, but between aproximately 10–15 daughter branches of Proto-Indo-European and a slightly lesser number of daughter branches of Proto-Uralic, this significantly increases the possibility of accidental similarities, mistaken for genuine cognacy.

On the other hand, this certainly does not explain the very fact of widely varying figures: clearly, 35% of matches between Indo-European and Altaic vs. 14% of matches between Indo-European and Sino-Tibetan is hard to interpret as a result of pure chance (with pure chance, the figures would be expected to be more comparable). The fact that S. Starostin's table lends itself easier to network-like rather than tree-like interpretation would, in my opinion, be easier to explain not as a result of accidence or even linguistic contacts, but as a direct consequence of UISD, operating randomly on all the daughter branches of the macrofamilies concerned, and driving all of the "cognacy" numbers up due to our inability (or, sometimes, unwillingness) to detect it.

Take, for instance, the comparison of Indo-European *ed- 'to eat' with Proto-Altaic *ite 'to eat'. The reconstructed words share the exact same basic meaning and obey regular phonetic correspondences, originally formulated by V. Illych-Svitych for Nostratic. However, while the Indo-European root is indeed Proto-Indo-European (Hittite ad-, Old Indian ad-, Latin ed-, Germanic *it-, etc.), the "Altaic" root is really only found in the required meaning in Mongolian (Proto-Mongolic *ide- 'to eat'). A much more realistic candidate for Common Altaic 'to eat' is the root *ǯē, reflected in Proto-Turkic (*yē- 'to eat'), Proto-Tungus-Manchu (*ǯe- 'to eat'), and Korean (Middle Korean čā- 'to eat'), further comparable with Proto-Fenno-Ugric *sewe or *seye 'to eat'. Distribution-wise, this match clearly outbets *ed- / *ite as the "optimal candidate" for the meaning 'to eat' on the "Nostratic" level of comparison.

So what exactly happened here? One possibility is random coincidence, which works for us if we do not accept the Nostratic hypothesis, but is less easy to believe if we do (an issue not fully relevant for the current discussion). Another option is "contact", which cannot be totally ruled out, but is highly unlikely, since borrowing of basic lexicon suggests intensive language contacts on a significant scale, which have certainly not been identified between Indo-European and Mongolic. What remains is UISD – in this case, an assumption that the Nostratic lexeme, ancestral to Indo-European and Mongolic, originally had a meaning that was *close* to 'eat' (for instance, 'bite', 'chew', 'gnaw', all of which may easily yield the more general meaning 'eat'), and effectuated the semantic transition independently in Proto-Indo-European and Proto-Mongolic, a process completely analogous to the one described earlier for *cluas* and *klautso*.

Obviously, at the level of our current elaboration of long-range relationship hypotheses such as Nostratic the suggested scenario is highly speculative. But, on the other hand, it hints at a way of resolving one of the most easily noticeable flaws of comparative reconstruction: the "rampant synonymy" that seems to plague published corpora of protolanguage etymologies, be it the classic etymological dictionaries of reliably established families (e.g. J. Pokorny's dictionary of Indo-European), or more controversial collections of higher-level etymologies (such as the recently published Nostratic dictionary by A. Dolgopolsky). Roughly speaking, if we see only one basic equivalent for such meanings as 'head', 'black', or 'drink' in historically attested languages, but five, six, or seven such equivalents postulated for their reconstructed common ancestor, this is more often than not the result of UISD, which the author of the corpus was not able to detect or, more likely, was not willing to, since strict semantic reconstruction rarely constitutes a top priority for the comparative linguist – a situation that, hopefully, will eventually change due to obvious progress in the field of diachronic semantic typology.

CONCLUSION

Although much has been said and written in the past fifty years about the allegedly fatal shortcomings of lexicostatistics, reality is different: some shortcomings have certainly been demonstrated, yet none of them have been shown to be genuinely "fatal". Maligned as it is, lexicostatistics has clearly stabilized its position in the average arsenal of comparative-historical methodology, since it is a reasonable, uniform, formalized, historically interpretable, and easily applicable procedure (not least of all, understandable for most historical linguists who are too busy studying language data to spend their time comparing models of various degrees of mathematical complexity). In fact, as historical linguistics becomes more and more involved in the sphere of interdisciplinary studies, interest in lexicostatistics seems to be growing, since no comparable quantitative alternative to the method has been found so far.

Our task, therefore, is to find out and explore as many of these "shortcomings" as possible – they represent valuable exceptions that may show exactly under which contexts lexicostatistical principles may be failing; an excellent example of this is the discovery of the crucial difference between internally and externally driven lexical replacement. On the other hand, it is also vital that lexicostatistical analysis (as well as any other formal statistical or probabilistic methods, for that matter) always go hand-in-hand with rigorous comparative research, because no conclusions about linguistic prehistory will ever be accepted by historical linguists if they cannot be aligned with a fully credible historical scenario.

Many of the misunderstandings and misgivings about lexicostatistics are due to the fact that the procedure, on the surface, may be seen as deceptively simple when, in fact, to yield results that are both generally credible and can be matched with a precise historical scenario, lexicostatistics has to be run with very strict attention to a whole number of details, preferably in "bootstrapping mode" where lexicostatistical conclusions and comparative procedures complement each other and correct each other's shortcomings. This approach is advocated in the construction of the Global Lexicostatistical Database (GLD), the latest project by the Moscow school of comparative linguistics, whose aim is not only to collect properly selected lexical items for the Swadesh wordlist from all of the world's languages, but also to detect and explore all of the different paths of the historical evolution of the basic lexicon – including the correlation of internally to externally driven change and the phenomenon of UISD.

So, if lexicostatistics is to be used "smartly", that is, in tight collaboration with classic comparative linguistics, will this mean that the results will lend themselves easier to tree-like or network-like interpretations? Essentially, all of the re-checkings of lexicostatistical findings by means of standard comparative linguistics that I have been advocating for earlier would lead to "disentangling the net" – helping us to identify the links between words that are due to non-genetic factors and remove them from the final structure. There is nothing wrong with using networks in historical linguistics; the important point is not to lose sight of the best possible tree behind the net, because if we do, lexicostatistics, and historical linguistics in general, lose one of their main points – the ability to correlate linguistic evidence with a plausible chronological scenario for the gradual spreading of the speakers of one proto-language over a particular "ethnolinguistic" area.

REFERENCES

Arapov; M.V. & M.M. Hertz (1974) Matematicheskije metody v istoricheskoj lingvistike [Mathematical methods in historical linguistics] (Moscow: Nauka Publishers).

Bateman, R., I. Goddard, R. O'Grady, V. Funk, R. Mooi, W. Kress & P. Cannell (1990) 'Speaking of forked tongues: The feasibility of reconciling human phylogeny and the history of language', Current Anthropology 31:1–24.

Bergsland, K. & H. Vogt (1962) 'On the Validity of Glottochronology', Current Anthropology 3: 115–53.

Blust, R. (2000) 'Why lexicostatistics doesn't work: the 'universal constant' hypothesis and the Austronesian languages', In McDonald Institute for Archaeological Research (ed), Time Depth in Historical Linguistics (Oxford: Publishing Press): 311–31.

Campbell, L. & W.J. Poser (2008) Language Classification: History and Method. (Cambridge: University Press).

Embleton, S. M. (1986) Statistics in Historical Linguistics (Bochum: Studienverlag Dr. N. Brockmeyer).

Gray, R.D. & Q.D. Atkinson (2003) 'Language-tree divergence times support the Anatolian theory of Indo-European origin', Nature 426: 435–9.

Haarmann, H. (1990) Language in its Cultural Embedding (Berlin: Mouton de Gruyter).

Haspelmath, M. & U. Tadmor (2009) 'The Loanword Typology Project and the World Loanword Database', in M. Haspelmath & U. Tadmor (eds), Loanwords in the World's Languages: A Comparative Handbook (The Hague: Mouton de Gruyter): 1–34.

Hoijer, H. (1956) 'Lexicostatistics: a critique', Language 32: 49–60.

Hymes, D. (1973) 'Lexicostatistics and glottochronology in the nineteenth century. With notes towards a general history', in I. Dyen (ed), Lexicostatistics in genetic linguistics. Proceedings of the Yale Conference, Yale University, April 3–4 1971 (The Hague: Mouton de Gruyter): 122–76.

Kassian, A.; Starostin, G.; Dybo, A. & V. Chernov (2010) 'The Swadesh wordlist: an attempt at semantic specification', Journal of Language Relationship 4: 46–89.

McMahon, A. & R. McMahon (2005) Language Classification by Numbers (Oxford: University Press).

Pagel, M.; Atkinson, Q.D. & A. Meade (2007) 'Frequency of word-use predicts rates of lexical evolution throughout Indo-European history', Nature 449 (7163): 717–20.

Peiros, I. (2000) 'Family diversity and time depth', in McDonald Institute for Archaeological Research (ed), Time Depth in Historical Linguistics (Oxford: Publishing Press): 75–108.

Renfrew, C.; McMahon, A. & L. Trask (eds) (2000) Time Depth in Historical Linguistics (Cambridge: The McDonald Institute for Archaeological Research).

Renfrew, C. & P. Forster (2006) Phylogenetic Methods and the Prehistory of Languages (Cambridge: The McDonald Institute for Archaeological Research).

Starostin, G. (2002) 'On the Genetic Affiliation of the Elamite Language', Mother Tongue VII: 147–70.

Starostin, G. (2008) 'Making a comparative linguist out of your computer: problems and achievements. Talk given at the Santa Fe Institute, August 12, 2008. Online version: http://starling.rinet.ru/Texts/computer.pdf.

Starostin, G. (2010) 'Preliminary lexicostatistics as a basis for language classification: a new approach', in Вопросы языкового родства (Journal of Language Relationship) 3: 79–117.

Starostin, Sergei A. (1989) 'Sravnitel'no-istoricheskoje jazykoznanije i leksikostatistika [Comparative-historical Linguistics and Lexicostatistics]', Linguistic reconstruction and the prehistory of the Ancient East (Moscow: Nauka): 3–39.

Starostin, Sergei A. (2000) 'Comparative-historical Linguistics and Lexicostatistics', in McDonald Institute for Archaeological Research, Time Depth in Historical Linguistics (Oxford: Publishing Press): 223–59. [english translation of S. Starostin 1989].

Starostin, S.A. (2003) 'Statistical Evaluation of the Lexical Proximity between the Main Linguistic Families of the Old World', Orientalia et Classica III: Studia Semitica. Moscow, RSUH Publishers: 464–84.

Starostin, S.A. (2007a) 'Computer-based simulation of the glottochronological process', in: S.A. Starostin, Raboty po jazykoznaniju [Works on Linguistics] (Moscow: Jazyki slav'anskikh kul'tur): 854–62.

Starostin, S.A. (2007b) 'Indo-European among other language families: problems of dating, contacts and genetic relationships', in S.A. Starostin, Raboty po jazykoznaniju [Works on Linguistics] (Moscow: Jazyki slav'anskikh kul'tur): 806–20.

Swadesh, M. (1952) 'Lexicostatistic dating of prehistoric ethnic contacts', Proceedings of the American Philosophical Society 96: 452–63.

Swadesh, M. (1955) 'Towards greater accuracy in lexicostatistic dating', International Journal of American Linguistics 21: 121–37.

Swadesh, M. (1965) 'Lingvističeskie sv'azi Ameriki i Evrazii' [Linguistic Ties Between America and Eurasia], Etimologija 1964 (Moscow: Nauka): 271–322.

Vovin, A. (2002) 'Building a 'bum-pa for Sino-Caucasian: a reply to Sergei Starostin's reply', Journal of Chinese Linguistics 30(1): 154–71.

ANALYZING DIALECTS BIOLOGICALLY

Jelena Prokić and John Nerbonne

1. INTRODUCTION

Dialectometry is a branch of linguistics whose main goal is the development and application of quantitative methods that enable researchers to explore relationships among dialects in an analytical way while taking into account large amounts of data. Most of the work done so far in dialectometry has focused on the differences between dialect varieties at the lexical and phonetic level, i.e., differences in vocabulary and pronunciation. However, there are projects that were concerned with the differences at the morphological (internal word structure) and syntactic level (concerned with structure in phrases and sentences). Regardless of the level at which the differences between the dialects are investigated, dialectometry has benefited from related developments in biology, especially those in population genetics and phylogenetics. These include use of sequence alignment techniques, hierarchical clustering, bootstrapping, and dimensionality reduction techniques, just to name a few.

This paper presents a line of research in dialectometry where the distances between the dialects are measured at the phonetic level. Application of quantitative methods to dialect pronunciation data consists of three major steps a) measuring distances, which in this paper will be done via string alignment / distance algorithms; b) detection of dialect groups; and c) linguistic interpretation. While any of the three major steps could be a subject of a separate survey paper, we focus here on the last step (c), which is concerned with identifying the linguistic basis for the automatic classification of dialects. We show how recent work that automatically identified characteristic *words* in given regions may easily be extended to allow the automatic identification of *sounds* characteristic of a given region. We also briefly present the first two steps for those less familiar with dialectometry, and we sketch some of the problems present in the quantification of dialect data (language data in general) using some of the mentioned methods. We first give short description of the data used throughout this paper.

2. DATA SET

In this paper we use Bulgarian dialect data that comes from the project *Buldialect – Measuring Linguistics Unity and Diversity in Europe*[1] to illustrate various methods used in dialectometry to measure and visualize the data. The Buldialect data set consists of the pronunciation of the 157 words collected at 197 villages distributed all over Bulgaria. Words included are frequent words that were collected from all, or almost all of the 197 sites. Regarding the choice of words, only words which are expected to show some degree of phonetic variation were included. There are in total 39 different dialectal features which have been represented in the chosen 157 words. The full list of 157 words and dialectal features present in these words can be found in Prokić et al. (2009) and Houtzagers, Nerbonne and Prokić (2010). Five words that have lower coverage than the rest of the words were excluded form all experiments presented in this paper, and the total number of words that we work with is 152.

3. STRING ALIGNMENT

The first step in the quantitative analyses of dialect phonetic variation is to measure the distances between various pronunciations in the data set. Hoppenbrouwers & Hoppenbrouwers (2001) address this problem by computing the differences of relative phone frequencies in various dialects (phones are individual sounds such as the 'p' sound in 'pail', or the 'l' sound). They also proposed a similar method based on the differences of the relative frequencies of different articulatory feature values of phones (features are properties, such as the property of being a vowel, or the property of being pronounced using an obstruction of the vocal path at the two lips, as in 'p'). Both of these frequency-based approaches do not take into account the ordering of the phones in a word. In order to make our measurements sensitive to the ordering of phones in a word, we must first align two pronunciations by means of the sequence (or string) alignment algorithms. We present two approaches to automatic string alignment used in dialectometry in the next two subsections.

3.1 Pairwise alignment

String alignment techniques have been introduced into dialectometry with the work of Brett Kessler who has used Levenshtein algorithm to calculate the pronunciation distance between the Irish Gaelic dialects (Kessler 1995). Application of the Levenshtein algorithm in dialectometry was later further developed and improved at the University of Groningen and applied to many languages in order to detect main dialect groups: Dutch (Nerbonne et al.1996; Heeringa 2004), Sardinian (Bolognesi

---

1   Volkswagen Foundation grant to P.I. Prof.Erhard Hinrichs, Tübingen.

and Heeringa 2002), Norwegian (Gooskens and Heeringa 2003), German (Nerbonne and Siedle 2005) and Bulgarian (Osenova et al. 2009).

The Levenshtein, or string edit distance, algorithm (Levenshtein 1966) is a dynamic programming algorithm used to measure the distance between two strings. The distance between two strings is defined as the smallest number of insertions, deletions and substitutions needed to transform one string to the other. We illustrate how one pronunciation of Bulgarian word *аз* 'I', namely [jɑ] (Aldomirovtsi) can be transformed into another [ɑs] (Asparuhovo):

$$
\begin{array}{ccc}
\text{j} & \text{ɑ} & \text{-} \\
\text{-} & \text{ɑ} & \text{s} \\
\hline
1 & & 1
\end{array}
$$

The minimal number of required operations is two: [j] has to be inserted/deleted in the word initial position, and [s] has to be inserted/deleted in the word final position. If the cost of each operation is 1, then the Levenshtein distance between these two strings is 2, and 2/3 if the distance is normalized by the length of the alignment. Treating the differences between phones in a binary fashion, i.e. same or not the same, is very simplistic model of sound change and for that reason very often unpopular among linguists. The cost of replacing one segment by another can be made more sensitive by basing it on articulatory features (Heeringa 2004) or automatically induced from the alignments (Prokić 2010; Wieling et al. 2012). The choice of the operation weights depends on the research goal. Whether or not one uses a segment weighting scheme leads to only minor differences in measurements at the aggregate level (Heeringa 2004). If one is interested in the more detailed analysis of the alignments, e.g. extraction of regular sound correspondences, then using a differential segment weighting produces more accurate alignments (Wieling et al. 2009) and is better suited for the dialect analysis at the segment level.

In order to calculate distances between each pair of sites in the data set, each pronunciation of a given word collected at one site is compared to the pronunciation of the same word at the other site by means of the Levenshtein algorithm. The distance between two sites is the mean of all word distances calculated for those two sites. The final result is a *site x site* distance matrix.

3.2 Multi-string alignment

Another approach to string alignment is multiple string alignment where all strings are aligned and compared at the same time. Automatic multiple string comparison is considered *the holy grail* of molecular biology (Gusfield 1997: 332). This type of string comparison, albeit executed manually, rather than automatically, has played a central role in linguistics ever since the late 19th century and the development of the comparative method of linguistic reconstruction (Campbell 2004). In the comparative method, identification of regular sound changes has played a major role in the identification of genetically related languages. The correct analysis of sound changes requires the simultaneous examination of corresponding sounds in the multiply aligned strings. Historical linguists align the sequences manually. In recent decade several algorithms for multiple string alignment in linguistics were developed (Bhargava and Kondrak 2009; Prokić et al. 2009; Steiner 2011; List 2012). In Prokić et al. (2009), the ALPHAMALIG algorithm was applied to dialect pronunciation data for the first time to multi-align word pronunciations. We illustrate the results of automatically aligning six pronunciations of word *аз* 'I':

j	ɑ	-	-	-	-
-	ɑ	s	-	-	-
j	ɑ	z	e	-	-
j	ɛ	-	-	-	-
j	ɑ	z	e	k	a
-	ɒ	s	-	-	-

The advantages of this type of alignment are twofold:
– First, it is easier to detect and process corresponding phones in words and their alternations (like [ɑ] and [ɛ] and [ɒ] in the above example).
– Multi-aligned strings, unlike pairwise aligned strings, contain information on the positions where phones are inserted or deleted in both strings. This leads to different distances between the strings as compared to the pairwise approach. In multi-aligned comparison the number of mismatching phones between [jɑ] and [ɑs] is 2/6 while it is only 2/3 if assayed based on the isolated pair (if distances are normalized).

Evaluation of the alignments automatically produced by ALPHAMALIG has shown that it is over 93% correct when compared to a manually corrected "gold standard" (Prokic et al. 2009) on the Buldialect data set.

The distances between the aligned strings can be calculated by counting the number of mismatching positions in a binary fashion or using some of the weighting schemes mentioned above.

4. DETECTION OF GROUPS

Once the distances between each pair of sites (villages) have been calculated, groups of dialects and their relatedness have to be reconstructed based on the estimated distances. Below we mention some of the methods most frequently used in dialectometry.

4.1 Clustering

A distance matrix that contains information on the distances between each pair of villages in the data set can be analyzed using clustering techniques, and later projected onto a map to check the geographical distribution of the groups obtained. Hierarchical clustering techniques were introduced into dialectometry by Hans Goebel (1982; 1983) who was the first to use clustering in analyses of dialect variation. He performed cluster analysis to detect the most important dialect groups and show their geographical spread by coloring groups detected by clustering differently on the map. Ever since, clustering has been commonly used to group dialects and analyze their relationship. In Figure 1 we present the dendrogram and the projection of the detected groups on the map of Bulgaria generated using Gabmap dialectometry software (Nerbonne et al. 2011) developed at the University of Groningen.[2]

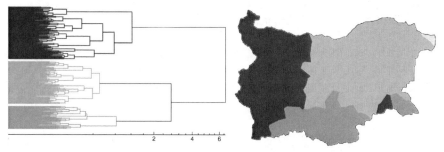

Figure 1: Dialect groups in Bulgaria identified using Ward's clustering method.

However, clustering techniques produce unstable results meaning that very small differences in the input matrix can lead to very different groupings of the data (Jain et al. 1988). In biology, in order to obtain stable clustering results a bootstrap procedure is often employed by randomly resampling the observed data (Felsenstein 2004). In dialectometry, Nerbonne et al. (2008) introduced *noisy* or *composite clustering,* in which small amounts of random noise are added to the matrices during repeated clustering. Tested on dialect data, bootstrapping and noisy clustering produce distance matrices that correlate nearly perfectly (*r = 0.997*). Unlike bootstrapping, noisy clustering can be applied on a single distance matrix which makes it easily applicable in dialectometry if distances between the sites are represented by a *site x site* distance matrix.

4.2 Network representation

There is a problem with using hierarchical clustering to determine the historical relationship among dialects, namely that this approach assumes an underlying tree model of dialect change. The relations between the groups produced by hierarchical

clustering are frequently represented by a bifurcating tree diagram called dendrogram (as shown in Figure 1). This representation of language relatedness suggests
that the innovations occur exclusively in the process of transmission from a mother
language variety to daughter varieties. Just as in biology, bifurcating phylogenetic
trees are used to model acquisition by inheritance only. Already in the 19th century,
Johannes Schmidt (1872) argued that innovations in languages are spread through
borrowing, i.e. he argued for the non-hierarchical diffusion of linguistic innovations from multiple sources. Borrowings that occur between languages correspond
to lateral transfer in biology, and they cannot be modelled using tree representation.
In order to visualize evolutionary relationships that include lateral tranfer, biologists use phylogenetic networks. One of the most popular method for reconstructing
phylogentic networks is Neighbor-Net, available as part of the Splits Tree software
(Huson and Bryant 2006).[3] In the past ten years there have been an increasing number of studies in linguistics that use this method to infer and visalize the relationships between language varieties. One important property of the Neighbor-Net algorithm is that, if the input distances are circular, it will return the collection of
circular splits, i.e. the network. If the input distances are additive, on the other
hand, it will return the corresponding tree (Bryant and Moulton 2004). This property enables researchers to see if the data is tree-like or network-like. In Figure 2 we
use Neighbor-Net to analyze the same distance matrix used to produce dendrogram
in Figure 1. By visually inspecting the network, we can identify three groups in the

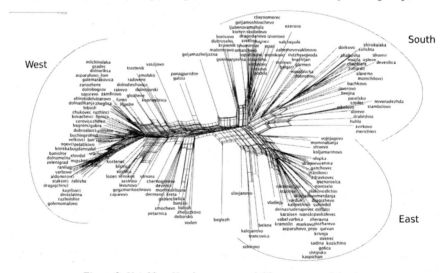

Figure 2: Neighbor-Net detects network-like structure of the data.

data, namely the West, East and South. However, the network representation allows
us to see that there are many conflicting signals represented as reticulations (lines
connecting radial branches), which makes the data look more network-like than

3   http://www.splitstree.org/

tree-like. This is by no means a surprising result, since dialects often form a continuum rather than groups of clearly separated varieties (Chambers and Trudgill 1998). The innovations are spread through borrowing and extensive social contact. For that reason, networks are more realistic representations of the relations between dialect varieties. Unfortunately, there is no direct way to link this kind of representation and geographic data, i.e. to project data onto the map, which is very important element of the research in traditional dialectology and in dialectometry as well. Another method frequently used in biology, namely multidimensional scaling, allows us to represent dialect variation as a continuum and project the results on the map.

4.3 Multidimensional scaling

Multidimensional scaling is a dimensionality-reducing technique used in exploratory data analysis and a data visualization method, often used to look for separations of data groups (Legendre and Legendre 1998). It analyses the set of distances between elements and attempts to arrange elements in a space within a certain small number of dimensions, which, however, accord with the observed distances. It was used for the first time in linguistics by Black (1973) and in dialectology by Embleton (1993). The plot of the first two extracted MDS dimensions obtained by applying multidimensional scaling to our distance matrix is presented in Figure 3, where

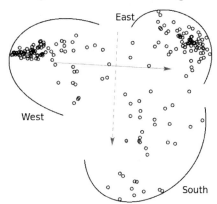

Figure 3: MDS-plot of the first two extracted dimensions.

the MDS plot shows two relatively homogeneous groups of varieties, West and East, and a third heterogeneous group that includes varieties from the South of Bulgaria.

Nerbonne and Heeringa (1998) were the first to project the results of MDS on a map by extracting the first 3 MDS dimension and associating each dimension with a color (red, blue and green). Each village in a data set was represented as a mix of these 3 colors depending on its coordinates in the MDS analysis. The space between the sites was colored by interpolation. The results of this technique applied on a Buldialect data set is shown in Figure 4 (limited to grey tones in print).

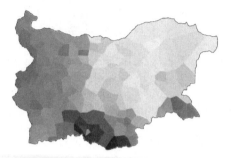

Figure 4: First 3 MDS dimensions projected on a map of Bulgaria.

This visualization technique enables us to detect three main dialect groups and at the same time to portray the degree of their linguistic heterogeneity (spread). It is especially suitable for the data that forms a continuum, like dialect data, rather than clearly separated groups.

5. CLUSTER DETERMINANTS

Most of work done in dialectometry so far has been focused on the first two steps: calculation of distances on the aggregate level and detection of dialect groups by means of some of the described methods. Settling on an appropriate linguistic interpretation of an aggregate analysis has always been considered the main drawback of dialectometry and made it less popular among more traditionally oriented dialectologists, who are often not as interested in the aggregate relations among sites, as in the concrete linguistic features that make one dialect distinct from other areas. Previous work in this direction include Nerbonne (2006), Grieve (2009) and Wieling and Nerbonne (2011). In this paper we present a method recently developed by Prokić, Çöltekin and Nerbonne (2012) that proceeds from a group of sites and identifies characteristic features of candidate dialect areas.

5.1 Method

The method proposed by Prokić et al. (2012) is general in that it can be applied to both numerical and to categorical data, requiring only that there be a numerical measure of difference defined for the data. It starts with data where the sites have already been split into groups, and it does not require any information on how the groups were obtained. This makes this method very general and also easily applicable in dialectometry.

The method proposed seeks the features which differ very little within the group in question and a great deal outside that group. It examines one candidate group g at a time that consist of $|g|$ sites among a larger area of interest G consisting of $|G|$ sites. This larger area G includes sites s within the cluster of interest and also those outside the cluster of interest g. The method assumes a measure of difference d between sites, always with respect to a given feature f.

A mean difference with respect to f is calculated within the group in question:

$$\overline{d}_f^g = \frac{2}{|g|^2 - |g|} \sum_{s,s' \in g} d_f(s, s')$$

and also involving elements outside the group in question:

$$\overline{d}_f^{\neg g} = \frac{1}{|g|(|G| - |g|)} \sum_{s \in g, s' \in \neg g} d_f(s, s')$$

where '$\neg g$' denotes the complement of g with respect to G.

Characteristic features are those with relatively large differences between $\overline{d}_f^g$ and $\overline{d}_f^{\neg g}$. The values obtained are sensitive to the size of the group under examination and the number of elements compared, which can be affected by missing data. Most importantly, the feature differences may systematically be influenced by differences in the natural variability of the data. For example, it appears that vowels are naturally more variable than consonants. To abstract away from this last influence, both $\overline{d}_f^g$ and $\overline{d}_f^{\neg g}$ are standardized by calculating the difference between z-scores. The mean and standard deviation of the difference values are estimated from all distance values calculated with respect to feature f. As a result the following measure is used:

$$\frac{\overline{d}_f^{\neg g} - \overline{d}_f}{sd(d_f)} - \frac{\overline{d}_f^g - \overline{d}_f}{sd(d_f)}$$

where d_f represents all distance values with respect to f. The scores are normalized for each feature separately.

The Buldialect set is blessedly complete, with data missing for very few pronunciations at very few sites. This means that we need not ask ourselves how often a given feature must be instantiated in a given region before we are willing to ask whether it might be characteristic. Prokić et al. (2012) discuss this problem.

5.2 Experimental setup

The method described is tested on the Buldialect pronunciation data (Section 2). Pronunciations of the 152 words from this data set were multi-aligned using AL-PHAMALIG algorithm. The automatically obtained alignments were very accurate, with scores ranging between 93 and 97 per cent depending on the evaluation method. We manually post-processed the alignments since we are primarily interested in the performance of the 'cluster determinants' method. However, because of the good quality of the automatically generated alignments the post-processing step could be avoided in future research. By proceeding from the multi-aligned data we assure that every position within a word is treated as a separate feature f. This is, incidentally, the point at which the present paper extends the work in Prokić et al. (2012).

The distances between each two sites were calculated by comparing the phones in each position in the multi-aligned pronunciations and taking the average of all obtained distances. The phones were compared based on the following weighting scheme: same phones have distance 0, same phones with different diacritics have distance 0.5 and different phones have distance 1. We use Gabmap software to do all calculations and obtain a *site × site* distance matrix.

In order to determine the optimal number of dialect groups in the data, we analyzed the distance matrix by means of MDS and Neighbor-Net (shown in Figures 3 and 4) which both revealed 3 relatively distinct groups. We tested several hierarchical clustering algorithms on our data by coloring the points in the MDS plot and additionally representing them with different symbols according to the 3-way divisions suggested by each of the clustering algorithms. In Figure 5 we present the 3-way division detected by Ward's algorithm, which we have found to be optimal for this data. The dialect regions detected can be seen on the map in Figure 1. These results agree with the traditional Bulgarian dialectology (Stoykov 2004) and quantitative analyses of the Buldialect data set (Houtzagers et al. 2010), which both distinguish western, eastern and southern dialects as the most important dialect groups.

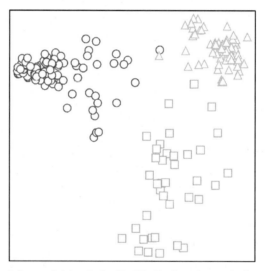

Figure 5: 3-way division derived by Ward's clustering method projected on MDS plot. Note that the colors correspond with those in the map.

In the final step, we apply the cluster determinants method described in section 5.1 in order to determine the linguistic bases of this 3-way division. Since our input data is multi-aligned, the features *f* that we are trying to recover using the described method are single positions in words. This is a step beyond Prokić, Çöltekin and Nerbonne (2012), which sought *words* characteristic of a given region. We now apply the same technique to seek characteristic *sounds*. In the next section we present the results.

5.3 Results

For each of the three dialect groups we calculated the most important linguistic feature, i.e. cluster determinants. In Table 1 we present the top five determinants for the western dialects. Each feature, i.e. word position, is presented within the word it occurs and marked in bold. We also present the standard pronunciation of the word in question.

Table 1: The five most important determinants for western dialects. We also give the word in which the feature appears and mark the feature itself with bold font.

Determinants	In cluster	Outside cluster
m lʲ a **k o** t o	o	u ʊ
bʲ a x m e	b	bʲ
dʒ **o** b	e	o i u a ʊ ɤ ɪ ʌ
nʲ a m a	n	nʲ
d ɤ n **o**	o	u ʊ

In Figure 6 we present the distribution of the first phone [o] from word мляко́то /mlʲakoto/ 'milk' (shaded dark on map) that was the highest scoring feature for the western dialects:

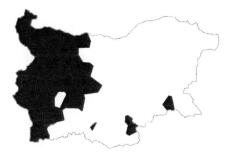

Figure 6: Distribution of the phone [o] in the second syllable of /mlʲakoto/, which clearly separates the western areas from the rest of the country.

The map in Figure 6 clearly shows that in the word мляко́то /mlʲakoto/ 'milk', the first /o/ (the vowel in the second syllable) is always realized as [o] in the west and as [u] or [ʊ] in the rest of the country. For that reason the distances among the varieties in the west with respect to this feature are very small when compared to the distances between those same varieties and varieties elsewhere.

In Table 2 we present five most important determinants for the eastern dialects:

Table 2: The five most important determinants for eastern dialects. The second one represents elision of [j].

Determinants	In cluster	Outside cluster
ts^j a l	ts^j	ts
- a z	-	j
g r o z d e	i	e ə ɣ ɪ
e d n o	i	e ə ɣ ɪ ɑ
d ɣ n o	o	u ʊ

In Figure 7 we show the distributional map for the most important cluster determinant for eastern dialects, realization of /ts^j/ in word цял /ts^{j}al/ 'whole'. In the east, it is always realized as [ts^j] and in the west and south as [ts].

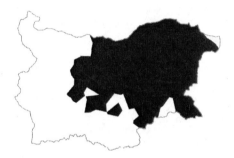

*Figure 7: The segment /ts^j/ is realized as [ts^j]
in the east and as [ts] in the west and south.*

The list of five most important determinants for the southern dialects is given in Table 3:

Table 3: The five most important determinants for southern dialects.

Determinants	In cluster	Outside cluster
tʃ e t ɣ	ə	a e i ɑ ɣ
r ʊ ts e	i ɨ	e ɑ ə ɛ
t o v a	-	i o u ɑ ɣ ʊ
d e r a	ə	a e ɑ ɣ ʊ
r ɣ ts e	c c^j k	ts ts^j

In Figure 8 we present the distribution of the most important determinant for the southern dialects. In word чета /tʃetɣ/ 'read – 1st sg', the segment /ɣ/ has realization [ə] in the south and [a], [e], [i], [ɑ], and [ɣ] in the rest of the country.

Figure 8: Segment /ɤ/ in word чема /tʃetɤ/ 'read –
1ˢᵗ sg' is realized as [ə] in the south.

The results presented suggest that this method is successful in recovering the most distinctive features for the area in question. In this paper, we have used multi-aligned data as input and treated each position in a word as a feature. However, this method can have any type of data as input, as long as the distances can be quantified. For example, Prokić et al. (2012) use whole words as features and quantify the distances between them using Levenshtein algorithm.

CONCLUSIONS

In this paper we have presented number of techniques taken from biology that are now standard tools in dialectometry, which is primarily concerned with measuring the distances between dialect varieties and their classification. Although biological and linguistic data differ to a great extent, techniques taken from biology have proven valuable in language data analyses. They have enabled us to analyze large amounts of data and overcome some of the methodological problems in earlier dialectology, which focused on identifying distinguishing individual features. Advances in the field of dialectometry have not been generally accepted by traditional dialectologists, perhaps because aggregate dialectometric analyses offer too little insight into the details they have focused on. In this work we have tested a new method that can overcome this problem, analyzing large amounts of data while at the same time preserving and sharpening a view on the linguistic details. This method can also be applied in other branches of linguistics that deal with quantitative language comparison. Clearly a great deal remains for future work. The technique should be applied to more data sets to gather more insight into its strengths and weaknesses, exposing further how it works and how it might be improved. One example of a point where a wider range of data must be examined is the parameter specifying how often a feature must be instantiated in a given region if it is to qualify at all as being "characteristic".

The present paper has examined specific positions (sounds) in specific words in an effort to find characteristic elements (the vowel in the second syllable of млякото /mlʲakoto/ 'milk') for a given cluster, while at the same keeps track of the context in which the element occurs. A great deal of linguistic interest is attached to the question of regular segment correspondences with respect to generally character-ized contexts (the /o/:/u/ correspondence in unstressed syllables) and we hope that the present paper has taken a step in that direction.

Finally, we should prefer to evaluate the work with respect to some independ-ent criterion, perhaps the reactions of dialect speakers (positive or negative) to given correspondences, or perhaps to their characterizations of the one or the other variant as like their own variety, or as rather different.

REFERENCES

Bhargava, A. & G. Kondrak (2009) 'Multiple Word Alignment with Profile Hidden Markov Mod-els.', in NAACL-HLT, Proceedings of Human Language Technologies: The Annual Confer-ence of the North American Chapter of the Association for Computational Linguistics, Com-panion Volume: Student Research Workshop and Doctoral Consortium (Boulder): 43–8.

Black, P. (1973) 'Multidimensional scaling applied to linguistic relationships', in Cahiers de l'Institut de Linguistique Louvain, Volume 3 (Montreal), Expanded version of a paper pre-sented at the Conference on Lexicostatistics (University of Montreal): 13–92.

Bolognesi, R. & W. Heeringa (2002) 'De invloed van dominante talen op het lexicon en de fonologie van Sardische dialecten', Gramma/TTT: tijdschrift voor taalwetenschap, 9(1): 45–84.

Bryant, D. & V. Moulton (2004) 'Neighbornet: an agglomerative algorithm for the construction of plenar phylogenetic networks', Molecular Biology and Evolution, 21: 255–65.

Campbell, L. (2004) Historical Linguistics: An Introduction. 2nd edition (Edinburgh: University Press).

Chambers, J.K. & P. Trudgill (1998) Dialectology (Cambridge: University Press).

Embleton, S. (1993) 'Multidimensional Scaling as a Dialectometrical Technique: Outline of a Re-search Project', in R. Köhler & B. Rieger (eds), Contributions to Quantitative Linguistics (Dordrecht: Kluwer): 267–76.

Felsenstein, J. (2004) Inferring Phylogenies (Sunderland: Sinauer Associates).

Goebl, H. (1982) 'Ansätze zu einer computativen Dialektometrie', in W. Besch, U. Knoop, W. Putschke & H.E. Wiegand (eds), Ein Handbuch zur deutschen und allgemeinen Dialekt-forschung. Handbücher zur Sprach- und Kommunikationswissenschaft, Volume I (Berlin/New York: de Gruyter Mouton): 778–92.

Goebl, H. (1983) '"Stammbaum" und "Welle". Vergleichende Betrachtungen aus numerisch-taxon-omischer Sicht.', Zeitschrift für Sprachwissenschaft 2, 3–44.

Grieve J. (2009) A Corpus-Based Regional Dialect Survey of Grammatical Variation in Written Standard American English, Ph.D. Dissertation (Flagstaff: Northern Arizona University).

Gusfield, D. (1997) Algorithms on Strings, Trees and Sequences: Computer Science and Computa-tional Biology (Cambridge: University Press).

Heeringa, W. & C. Gooskens (2003) 'Norwegian dialects examined perceptually and acoustically.', in J. Nerbonne & W. Kretzschmar (eds), Computers and the Humanities 37 (3) (Dordrecht: Kluwer Academic Publishers): 293–315.

Heeringa, W. (2004) Measuring dialect pronunciation differences using Levenshtein distance, Ph.D. dissertation (University of Groningen).

Hoppenbrouwers, C. & G. Hoppenbrouwers (2001) De indeling van de Nederlands streektalen: dia-lecten van 156 steden en dorpen geklasseerd volgens de FFM (Assen: Koninklijke Van Gor-cum).

Houtzagers, P.; Nerbonne, J. & J. Prokić (2010) 'Quantitative and Traditional Classifications of Bulgarian Dialects Compared', Scando-Slavica 56 (2): 29–54.

Huson, D.H. & D. Bryant (2006) 'Application of phylogenetic networks in evolutionary studies', in Molecular Biology and Evolution 23(2): 254–67.

Jain, A.K. & R.C. Dubes (1988) Algorithms for Clustering Data (New Jersey: Prentice Hall PTR).

Kessler, B. (1995) 'Computational dialectology in Irish Gaelic', in Proceedings of the European ACL (Dublin: Association for Computational Linguistics): 60–67.

Legendre, P. & L. Legendre (1998) Numerical Ecology, 2nd edition (Amsterdam: Elsevier).

Levenshtein, V.I. (1966) 'Binary codes capable of correcting insertions, deletions and reversals', Cybernetics and Control Theory 10(8): 707–10 [Russian orig. in Doklady Akademii Nauk SSR 163(4), 1965: 845–8].

List, J.-M. (2012) 'Multiple sequence alignment in historical linguistics. A sound class based approach', Proceedings of ConSOLE XIX, 242–60.

Nerbonne, J. (2006) 'Identifying Linguistic Structure in Aggregate Comparison', Literary and Linguistic Computing 21(4): 463–76.

Nerbonne, J. & W. Kretzschmar, Jr. (eds) (2006) 'Progress in Dialectometry: Toward Explanation', Literary and Linguistic Computing, 21(4): 387–98.

Nerbonne, J.; Heeringa, W.; Hout, E. van den; Kooi, P. van de; Otten, S. & W. van de Vis (1996) 'Phonetic Distances between Dutch Dialects', in G. Durieux, W. Daelemans & S. Gillis (eds), CLIN VI: Proc. of the Sixth CLIN Meeting (Antwerpen, Centre for Dutch Language and Speech): 185–202.

Nerbonne, J. & W. Heeringa (1998) 'Computationele vergelijking and classificatie van dialecten.', in Taal en Tongval; Tijdschrift voor Dialectologie 50(2): 164–93.

Nerbonne, J. & C. Siedle (2005) 'Dialektklassifikation auf der Grundlage aggregierter Ausspracheunterschiede', Zeitschrift für Dialektologie und Linguistik 72(2): 129–47.

Nerbonne, J.; Kleiweg, P.; Heeringa, W. & F. Manni (2008) 'Projecting Dialect Differences to Geography: Bootstrap Clustering vs. Noisy Clustering', in C. Preisach, L. Schmidt-Thieme, H. Burkhardt & R. Decker (eds) Data Analysis, Machine Learning, and Applications. Proc. of the 31st Annual Meeting of the German Classification Society (Berlin: Springer) 647–54.

Nerbonne, J.; Colen, R.; Gooskens, C.; Kleiweg, P. & T. Leinonen (2011) 'Gabmap – A Web Application for Dialectology.', Dialectologia. Special Issue II, 2011: 65–89.

Osenova, P.; Heeringa, W. & J. Nerbonne (2009) 'A Quantitative Analysis of Bulgarian Dialect Pronunciation', Zeitschrift für slavische Philologie 66(2): 425–58.

Prokić, J.; Wieling, M. & J. Nerbonne (2009) 'Multiple string alignments in linguistics.', in L. Borin & P. Landvai (chairs) Language Technology and Resources for Cultural Heritage, Social Sciences, Humanities, and Education (LaTeCH – SHELT&R 2009) EACL Workshop. (Athens): 18–25.

Prokić, J.; Nerbonne, J.; Zhobov, V.; Osenova, P.; Simov, K.; Zastrow, T. & E. Hinrichs (2009) 'The computational analysis of Bulgarian dialect pronunciation', Serdica Journal of Computing 3: 269–298.

Prokić, J. (2010) Families and Resemblances, PhD thesis (University of Groningen).

Prokić, J.; Coltekin, C. & J. Nerbonne (2012) 'Detecting Shibboleths.', in Proceedings of the EACL 2012 Joint Workshop of LINGVIS & UNCLH (Avignon).

Schmidt J. (1872) Die Verwandtschaftsverhältnisse der Indogermanischen Sprachen (Weimar: Böhlau).

Steiner, L.; Stadler, P. & M. Cysouw (2011) 'A Pipeline for Computational Historical Linguistics', Language Dynamics and Change 1(1).

Stoykov, S. (2004) Bulgarska dialektologiya, 4th ed. (Sofia).

Wieling, M.; Margarethe, E. & J. Nerbonne (2012) 'Inducing a Measure of Phonetic Similarity from Pronunciation Variation', Journal of Phonetics 40(2): 307–14. DOI: http://dx.doi.org/10.1016/j.wocn.2011.12.004

Wieling, M. & J. Nerbonne (2011) 'Bipartite spectral graph partitioning for clustering dialect varieties and detecting their linguistic features', in Computer Speech and Language 25: 700–15. DOI:10.1016/j.csl.2010.05.004.

Wieling, M.; Prokić, J. & J. Nerbonne (2009) 'Evaluating the pairwise string alignment of pronunciations', in L. Borin & P. Landvai (chairs), Language Technology and Resources for Cultural Heritage, Social Sciences, Humanities, and Education (LaTeCH – SHELT&R 2009) EACL Workshop. (Athens): 26–34.

RECONSTRUCTING THE LATERAL COMPONENT OF LANGUAGE HISTORY AND GENOME EVOLUTION USING NETWORK APPROACHES

Shijulal Nelson-Sathi, Ovidiu Popa, Johann-Mattis List, Hans Geisler,
William F. Martin and Tal Dagan

PREFACE

Genome evolution and the history of language development share many features. Both processes involve basic elements – words or genes – whose properties can change over time. An alteration of an element's property can lead to a change in its function that in turn may affect the structure and composition of the whole domain, be it a language or a genome. Similarly to genomes that owe their existence to their corresponding species, languages also exist as long as there exists a population of native-speakers. Both genomes and languages may vary within the population. Eventually, the population that carries the domains – speakers or organisms – may split and continue to change independently, resulting in a divergence event and the origin of new languages or species.

First investigations into the similarities between language and species evolution are documented in the early modern period. These were the times when "Catastrophism", the leading paradigm of natural history, linguistics, and geology during the Middle Ages and the Early Modern Period, lost its power (Wells 1973). Under the slogan, "The present is the key to the past," which was originally coined by geologists (Cannon 1960), genealogical relations were inferred for species and languages. Shared traits in their present form were interpreted as evidence for their past identity, and the family tree became the leading metaphor to describe genealogical relations in linguistics as well as in biology.

As Geisler & List (this volume) point out, methods for phylogenetic reconstruction were developed independently in biology and linguistics, with August Schleicher (1821–1861) being among the first linguists to model language evolution by means of bifurcating trees (Schleicher 1853a, Schleicher 1853b, see Figure 1), and Charles Darwin (1809–1882) being among the first biologists to illustrate the splitting of ancestor species into their descendants with help of the family tree schema (Darwin 1859, see Figure 2). Neither Darwin nor Schleicher were the first to use trees to depict species or language evolution, yet both made the tree model popular in their respective disciplines (Ragan 2009; Sutrop 1999).

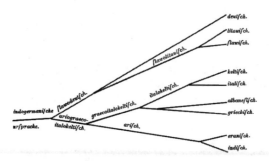

Figure 1: Family tree of the Indo-European languages
by August Schleicher (1861:7)

Biologists and linguists soon became aware of some striking similarities, not only between the objects, but also between the processes that they were investigating. Darwin briefly addressed the topic of language evolution in his work (Darwin 1859), and August Schleicher devoted an essay to the German biologist Ernst Haeckel (1834–1919). The essay, dealing with parallels and differences between language classification and species evolution, was published as an open letter (Schleicher 1863). In his essay, Schleicher addressed explicitly the importance of the uniformitarian principle (ibid. 10f) and the family tree model (ibid. 14f) in both disciplines, and emphasized the differences between the biological and linguistic entities (Schleicher 1863).

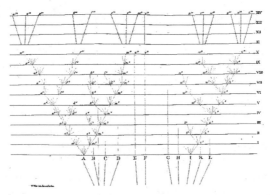

Figure 2: Charles Darwin's family tree (Darwin 1859:
illustration in the addendum)

LATERAL TRANSFER IN GENOME EVOLUTION AND LANGUAGE HISTORY

In biology, the family tree model remained the leading evolutionary model for a long time. In linguistics, however, scholars began quite early to question its adequacy to depict the complexity of language history in a realistic manner. As Geisler & List (this volume) emphasize, linguists have long recognized that historical relations between languages are not necessarily *vertical*, i.e. genealogical, but may also be *horizontal*, i.e. non-genealogical, resulting from language contact or lexical borrowing. This finds its reflection in the fact that not long after Schleicher popularized the use of family trees in linguistics, Johannes Schmidt (1843–1901) proposed his *Wave Theory*, as an alternative theory of language evolution. However, lacking the suggestive force of the tree metaphor, Schmidt's *Wave Theory* remained an impalpable concept, as can be seen from the many different attempts in the history of linguistics to visualize it properly (cf. Geisler & List this volume). For many years, linguists would use the family tree while at the same time criticizing its adequacy. Although many scholars followed Schmidt's example and emphasized the inadequacy of the family tree in linguistics, none of the many alternative models that were proposed, be it waves (Schmidt 1872; Hirt 1905), chains, or even animated pictures (Schuchardt 1870 [1900]), gained acceptance among all scholars. For many years, linguists would use the family tree while at the same time criticizing its adequacy.

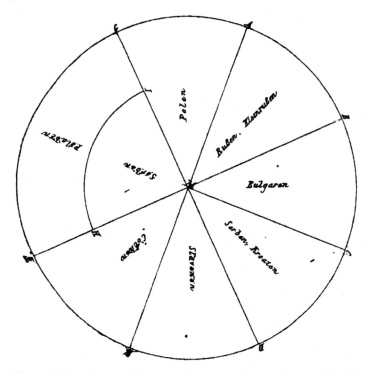

Figure 3: Schmidt's 'Wave Theory' in his own visualization (Schmidt 1875:199)

In contrast to the controversial character of the tree model in linguistics, the bifurcating trees were considered as the only reasonable model to describe species evolution for many decades. While this still holds for macroscopic evolutionary processes, studies in microbiology revealed that the tree model is insufficient for an adequate description of microbial evolution (Dickerson 1980; Doolittle 1999; Ochman 2000; Bapteste et al. 2009). Prokaryotes are capable of acquiring new genetic material from their neighbourhood or directly from the environment and incorporate it into their genomes in a process termed lateral gene transfer (LGT). Gene acquisition by lateral transfer in prokaryotes was first described in the 1950ies (Freeman 1951; Ochiai et al. 1959). The evolutionary implications of LGT bear strong resemblances to the process of borrowing during language evolution (Bryant et al. 2005; Pagel 2009). The development of advanced genome sequencing technologies enabled the investigation of microbial genomes at the DNA level, which led to the realization that LGT plays a major role in shaping microbial genomes (Koonin 2009; Bapteste et al. 2009; Popa and Dagan 2011).

Mechanisms of lateral gene transfer

Known mechanisms of lateral gene transfer include transformation, transduction, conjugation, and gene transfer agents (Thomas and Nielson 2005; Lang and Beatty 2007). Transformation involves the uptake of naked DNA from the environment. DNA uptake is enabled during a competence state that involves 20–50 proteins, including the type IV pillus and type II secretion system proteins (Chen and Dubnau. 2004; Thomas and Nielson 2005). In some species, an effective transformation requires the presence of uptake signal sequences (USSs). These are specific DNA motifs, about 10bp long, that are encoded within the recipient genome in a frequency that is much above that expected by random (Smith et al. 1995). Environmental DNA molecules bearing the USS motif are recognized by specific receptors at the cell surface, imported into the cytoplasm, and can then be readily integrated into the recipient chromosomes, usually via homologous recombination (Chen and Dubnau 2004).

Transduction is DNA acquisition following a phage infection. Phage recognise possible hosts by specific receptors found on the cell surface. Many phages include in their genomes chunks of DNA taken coincidentally from previous hosts. These are transferred to the new host during the integration of the phage genome into the host chromosomes. DNA integration into the host chromosome is generally mediated by the phage-encoded enzymes that specifically integrate the phage into the chromosome of the infected recipient (Thomas and Nielsen 2005; Lindell et al. 2004; Sullivan et al. 2006).

Conjugation is the transfer of DNA via plasmids, a process that is mediated by cell-to-cell junction and a tunnel through which the DNA is transferred. The transferred material is typically a plasmid that can pass through the tunnel during conjugation breaks off. Plasmids can integrate into the recipient chromosomes by homologous recombination that may entail insertion sequences (ISs) or other sequences conserved between plasmid and recipient chromosomes that carry the minimal sequence similarity required for homologous recombination (Chen et al. 2005).

Gene transfer agents (GTA) are phage-like DNA-vehicles that are produced by a donor cell and released to the environment. DNA stored in GTAs is imported into the recipient in a process that is similar to transduction. GTAs, unlike phages, are linked to transfer of genomic DNA only and they have no negative effect on the recipient. GTA systems have been documented mainly in oceanic a-proteobacteria, but also in few archaebacteria and some spirochaetes (Lang et al. 2012; Berglund et al. 2009; Zaho et al. 2010). A recent comparison of GTA-mediated gene transfer rates among various marine habitats revealed particularly high transfer rate in the open ocean, indicating the importance of this transfer mechanism for genome evolution in oceanic alpha-proteobacteria (Mcdaniel et al. 2010).

An additional transfer mechanism – nanotubes – was discovered recently (Dubey and Ben-Yehuda 2011). These are tubular protrusions composed of membrane components that can bridge between neighboring cells and conduct the transfer of DNA and proteins.

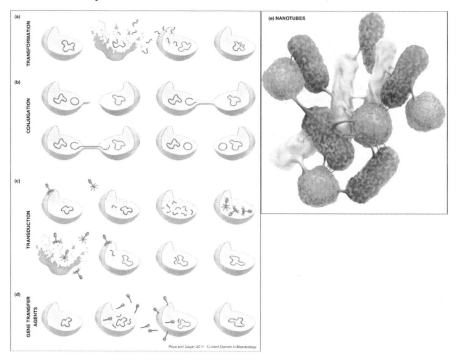

Figure 4: Lateral gene transfer mechanisms

Basic mechanisms of borrowing

Describing the character of languages in all their grammatical, phonetic, and lexical complexity is an extremely difficult task, even when disregarding their history. Consequently, the detailed study of language evolution is often reduced to the study of *lexical change*. The most common unit of the lexicon in a language is the word (or morpheme), a unit which is characterized by its *form*, and its *meaning* (similar to gene sequence and *function* in biology). While the form of a word is directly accessible and can be characterized as a sequence of sound segments, the meaning of normally polysemous words is far more difficult to describe. Like in biology, where protein function cannot be predicted from its sequence alone (in the lack of known homologs), no natural link between form and meaning (function) can be claimed for the linguistic sign. Any connection between form and meaning is only due to convention.

The basic processes of lexical change can be roughly divided into *vertical processes* resulting from semantic change or semantic innovations and *horizontal processes* resulting from borrowing. While vertical processes are *gradual*, horizontal processes are *discrete*, involving a *donor* and a *recipient* language. Evolutionary events affecting a single word can be roughly divided into those that change its form (*sound change*) and those that change its meaning (*semantic change*). Sound change is an overwhelmingly regular process (Hock and Joseph 1995: 241–278). Lexical change is defined as a change in the meaning of a sign compared to its ancestor while a change in the form resulting from regular sound change processes is disregarded (Gevaudan 2007: 14f). In a *direct transfer* both the form and meaning of a word are transferred as a whole from the donor to the recipient language. During *semantic transfer* (or *semantic borrowing*) a word in the donor language is reproduced in the recipient language by expanding the meaning of a given word in the recipient language to match the form-meaning unity in the donor language (*semantic transfer*, *semantic borrowing*, cf. Weinreich 1953: 48). For example, the standard Chinese *kāfēi* "coffee" was directly transferred from English, as is also evident from the similar pronunciation of the words. The standard Chinese *diàn* "electricity", on the other hand, has been indirectly transferred by extending the word's original meaning "lightning" (see Figure 5).

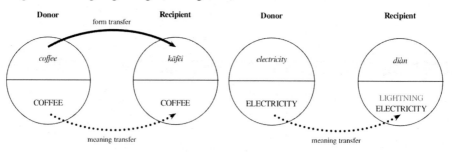

Figure 5: Direct transfer and reproduction

SIMILARITIES AND DIFFERENCES IN LANGUAGE AND GENOME EVOLUTION

The most striking difference between languages and genomes, as Geisler and List (this volume) point out, lies in the way they manifest themselves: genome evolution takes place in a materialistic world, while language history does not. Genes have a substance, words do not, and while the function of a gene is a product of its sequence, any connection between form and function of the linguistic sign is strictly arbitrary. These differences are reflected in the mechanisms that drive lateral gene transfer and lexical borrowing. The most striking of these differences is that the transfer of linguistic material is not restricted to the direct transfer of form. Lateral gene transfer only applies to the exchange of genetic material, i.e. the exchange of form as a vehicle of function. In contrary, lexical borrowing between languages can involve the transfer of both form and function, or the transfer of functions alone.

Lateral transfer frequency during genome evolution

Several experiments have been conducted in order to quantify the frequency of LGT in nature. For example, Babic et al. (2008) tested the success rate of gene acquisition by conjugation in *Escherichia coli*. Using a plasmid encoding a gene for fluorescence protein (YFP) they quantified the odds for a successful integration of plasmid genes into the recipient genome. They found that in 96% of the population the YFP gene was integrated into the chromosome and inherited to the next generation. The percolation of an acquired DNA within the population can be extremely fast in *Bacillus subtilis* where the cells are arranged in chains. Tracking the spread of an integrative and conjugative element (ICE) encoding a gene for green fluorescence protein (GFP) under the microscope showed that in 43 (81%) out of 53 cases a recipient cell turned into a donor and transconjugated the ICE to the next cell in line, often within 30 minutes (Babic et al. 2011).

Lateral gene transfer via transduction takes place during a phage infection. Hence gene acquisition by this transfer mechanism depends on the survival of the recipient. In a recent study Kenzaka et al. (2010) quantified the survival rate of phage infected enteric bacteria as 20% of the population. These surviving bacteria may acquire DNA from previous hosts of the attacking phage.

We know that LGT occurs in the laboratory, the issue is how often it occurs in the wild and how important it is during evolution. Phylogenetic reconstruction of microbial genes reveals that LGT plays a major role in shaping microbial genomes (Mirkin et al. 2003; Kunin et al. 2005; Dagan and Martin 2007; Halary et al. 2010; Kloesges et al. 2011). In a pioneering study, Lawrence and Ochman (1998) identified all *E. coli* genes that were acquired since its divergence from the *Salmonella* lineage by their aberrant codon usage. They estimated that 755 (18%) of the 4,288 genes in *E. coli* strain MG1655 were laterally acquired over a time period of about 14 million years (Myr) and estimated the LGT rate as 16Kb/1Myr per lineage (Lawrence and Ochman 1998). Using gene distribution patterns across 329 proteobacterial genomes, Kloesges et al. (2011) recently estimated that at least 75% of the protein families have been affected by LGT during evolution. Gene transfer rate in those families is on average 1.9 events per protein family per lifespan (Kloesges et al. 2011). Similar estimates were found in phylogenetic analyses of broader taxonomic samples (Mirkin et al. 2003; Kunin et al. 2005; Dagan and Martin 2007).

The impact of LGT during genome evolution can be estimated either by the proportion of recently transferred genes whose unusual base composition and codon usage still bears the marks of acquired DNA (Lawrence and Ochman 1998; Garcia-Vallve et al. 2000; Nakamura et al. 2004) or by phylogenetic analysis of individual genes including recent and ancient LGTs alike (e.g. Zhaxybayeva et al. 2006; Beiko et al. 2005; Puigbò et al. 2010; Chan et al. 2011). A survey of genes having aberrant nucleotide composition within proteobacterial genomes revealed that 21±9% of the genes in those genomes comprises recent acquisitions (Kloesges et al. 2011). Gene distribution patterns across the same species sample suggest that, on average, 74±11% of the genes in each genome have been laterally transferred at least once during evolution (Kloesges et al. 2011).

Lateral transfer frequency during language history

Borrowing frequency may vary dramatically during language evolution, depending on many different factors such as the sociocultural situation in which the respective language is used, the geographical distance of the language to other languages, or the prestige of specific language varieties within a given speech community. Borrowed vocabulary can affect only small parts of the lexicon of a given language (such as specific terms for cultural items), or alternatively result in a situation where large parts of the language lexicon are acquired or replaced and can be traced back to a donor language.

In the recently published World Loanword Database (WOLD, Haspelmath and Tadmor 2009) the frequency of direct borrowing events in a sample of 1460 glosses that were translated into 41 different languages was investigated. Borrowing rates in the database vary greatly, ranging from 1 % (Mandarin Chinese) to 62 % (Selice Romani) with an average of 25 % and a standard deviation of 13 % (Tadmor 2009; Figure 6)

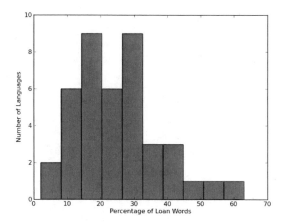

Figure 6: Histogram of borrowing frequencies in the WOLD database

PHYLOGENOMIC NETWORKS

Considering the frequency of lateral transfer events during the evolution of genomes and languages leads to the realisation that the tree model is missing a substantial portion of the evolutionary history in these domains. Studying the evolutionary dynamics of genomes and languages in detail requires alternative, more complex, models that allow to incorporate both vertical and horizontal transfers. Networks approach provides a more realistic model of microbial and language evolution than trees because they allow the reconstruction of non tree-like events such as recombination, gene fusion, and lateral gene transfer (Huson and Scornavacca 2011; Dagan 2011). Representing genealogical relations using a network is not new and has been documented even before Darwin's species tree was popularized (Ragan 2009). These earlier examples however focus on illustrating complex evolutionary relationship. The advance in statistical methods to analyse network properties enables the application of networks to evolutionary studies in a much more quantitative way (Dagan 2011).

A network constitutes a set of entities and the pairwise relations among them. The entities are termed *nodes* (or *vertices*) are connected by *edges* representation relationship (Newman 2010). Phylogenomic networks comprise completely sequenced genomes as nodes that are connected by edges of phylogenetic relations (Dagan 2011). Phylogenomic networks can be reconstructed from shared gene content (Dagan et al. 2008; Halary et al. 2010), shared sequence similarity (Lima-Mendez et al. 2008), or phylogenetic trees (Beiko et al. 2005; Popa et al. 2011). A more detailed phylogenomic network is the *directed network of lateral gene transfer* (dLGT) in which the nodes correspond to species or their ancestors and the directed edges represent recent lateral transfer events containing direction from donor to recipient as additional information (Popa et al. 2011).

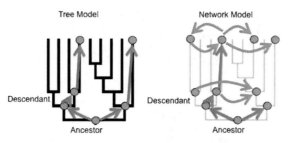

Figure 7: 'Tree Model' and 'Network Model'

Phylogenetic networks of languages

Linguists have long been aware of the problems that lexical borrowing poses to the tree model and tried to find more sophisticated ways to include the lateral component of language evolution in the evolutionary process representation. Given the need to model both vertical and horizontal processes, linguists would naturally turn to networks as a format to represent language evolution. Ignoring spurious hints to horizontal branches introduced into language trees that can be found in the literature rather early (Schuchard 1870 [1900]), the first explicit network approach can be found in a study by Bonfante (1931) where the complex relations between the Indo-European languages led the author to deny the possibility of true genealogical tree in this language family and propose a network model instead (see Figure 5 in Geisler & List this volume). Unfortunately, the use of networks by Bonfante and later similar studies (Southworth 1964, Anttila 1972) remained a mere visualization of the scholars' intuitions regarding patterns in the data and did not enable further insights regarding language evolution.

More recent attempts to reconstruct networks of language phylogeny include the applications of reticulated trees and split networks, which were initially developed in order to study genome evolution (Bryant 2005, McMahon 2005, Hamed and Wang 2006). While these methods can reveal the extent of non tree-like evolutionary dynamics, none of them can by used to estimate borrowing frequency. Thus, although based on quantitative data, split networks still remain a visualization tool.

Minimal lateral networks (MLN; Dagan et al. 2008), developed originally to study microbial genome evolution, enable an automatic inference of borrowing events in linguistic datasets and their visualization in a network form (Nelson-Sathi et al. 2011). The method is based on the construction of a phylogenetic network model in which presumed cognate words, i.e. words that go back to a common ancestor, are mapped on to a reference tree for the languages being analysed. Branching patterns in the data that are incompatible with the reference tree are explained by means of different borrowing models. These models allow for an increasing amount of borrowing events by which the patchiness of the data in comparison with the reference tree can be explained. Based on the assumption that the number of words that are used to express a certain set of concepts is approximately the same in all languages and throughout all times, a model is chosen which minimizes the difference in the average word inventories between the ancestor language and its descendants.

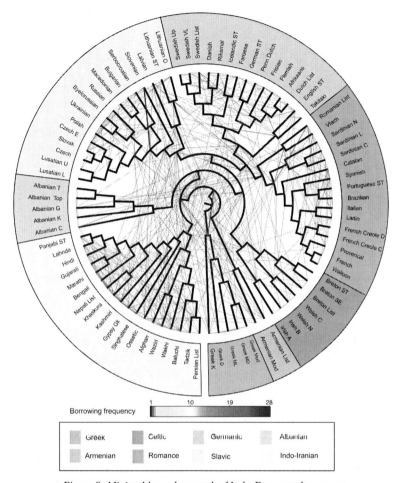

Figure 8: Minimal lateral network of Indo-European languages

Figure 8 shows a Minimal Lateral Network (MLN) of a dataset of 84 Indo-European languages (Dyen et al. 1992) representing both the vertical as well as the lateral components of language evolution reconstructed by the method. Nodes in this network represent contemporary (external) and ancestral (internal) languages. The ancestral languages correspond to the nodes in the reference tree. The vertices are connected either by branches of the reference tree, representing vertical inheritance, or by lateral edges, representing inferred borrowing events whose frequency corresponds to the edge width.

The method has several advantages over previously proposed ones: Since it infers concrete borrowing events, the results are transparent and can be directly tested against the data. Furthermore, additional analyses can be applied to the MLN addressing various questions such as what is the borrowing frequency observed in the data or which trends and barriers exist for borrowing dynamics during language evolution.

The benefits of network approaches in biology and linguistics

Phylogenetic networks – in contrast to phylogenetic trees – have many advantages when studying genome evolution or language history. They are more informative than simple family trees, since they do not ignore horizontal relations between genomes and languages. In contrast to the different Wave models, which were presented as alternatives to the tree model in historical linguistics (see Geisler and List this volume), phylogenetic networks do not result in a static visualization of relations between taxa, but provide a dynamic model of evolution in both linguistics and biology.

TRENDS AND BARRIERS

For a long time, the lateral components of genome evolution and language history have been ignored by biologists and linguists. Reticulated evolutionary events such as lateral transfer are often seen as irregular, chaotic events that blur the "real" phylogeny in both fields. The ability to include the lateral component of genome evolution and language history as an integral part of phylogenetic studies is, however, expected to promote our understanding of the trends and the barriers that underlay LGT and lexical borrowing dynamics. Experimental work shows that gene acquisition by LGT among prokaryotes is frequent and that the percolation of acquired DNA among populations and across generations is rapid. Phylogenomic analyses reveal that LGT has a substantial impact on long-term genome evolution, supplying a mechanism for natural variation that is specific for the prokaryotic domains and allows their adaptation in dynamic environments. Prokaryote genome evolution comprises thus vertical (tree-like) and lateral (network-like) components. At the same time, different types of barriers to LGT on the genomic, species, and habitat levels are becoming increasingly apparent (Popa and Dagan 2011).

Trends and barriers to lateral gene transfer

The dLGT network reveals clusters of densely connected donors and recipients that are very similar in their genomic nucleotide content (GC content) (Popa and Dagan 2011). The difference in genomic GC content between donors and recipients is <5% for most (86%) of connected pairs (Popa et al 2011). Furthermore, donor-recipient genome sequence similarity and LGT frequency are positively correlated (rs = 0.55, P << 0.01) (Popa et al. 2011). This suggests that LGT is more frequent among closely related species, having similar genomes, while LGT between distantly related species is more rare bring out the existence of a donor-recipient *similarity barrier*.

Microbes tend to delete non- functional or otherwise unneeded DNA from their genomes (Moran et al 2009, Burke and Moran 2011). Therefore, the fixation of the acquired DNA within the genome is highly dependent on its functionality or utility to the recipient under selectable environmental conditions (Hao and Golding 2006; Pal et al. 2005; Zhaxybayeva and Doolittle 2011). Most laterally transferred genes perform *metabolic functions,* e.g. they are responsible for harvesting energy or for the construction of cell components like amino- or nucleic acids, while the transfer of genes performing information processing (including replication, transcription, and translation) is rare (Popa et al 2011, Jain et al. 1999, Coscolla et al 2011). According to the complexity hypothesis (Jain et al. 1999), the scarcity of lateral transfer of information processing genes is attributed to their role in complex structures. Proteins that function in a complex structure, for example ribosomal proteins, are adapted to their common function. An LGT event that leads to replacement of such a gene with a less adapted homolog will result in a 'squeaking wheel' within the complex and reduced fitness of the recipient (Jain et al. 1999). This suggests that there is *a functional barrier* for LGT. Genomic fragments whose cloning into E. coli is lethal are suspects for encoding proteins whose acquisition in E. coli is extremely disadvantageous (Sorek et al. 2007). An extensive dataset of lethal fragments collected during genome sequencing projects of 79 diverse species showed that these fragments typically encode for single copy genes. The integration of an additional gene copy into the E. coli genome resulted in an elevated protein production that was lethal to the cell (Sorek et al. 2007) suggesting protein dosage as another functional barrier to LGT.

The physical distance between the donor and recipient in the LGT event depends upon the LGT mechanism (Popa and Dagan 2011). In transformation the distance between the donor and recipient depends upon the raw DNA stability within the environment (Majewski J, 2001). Conjugation requires that the donor and recipient will be close enough for the formation of the conjugation tunnel. Transduction is considered as the longest range LGT mechanism because it entails phage mobility (Majewski J, 2001). This suggests that most transfers should occur within habitats. The dLGT network reveals that indeed most (74%) of the detected LGT in the network occur between donors and recipients residing in the same habitat (Popa and Dagan 2011), indicating the presence of an *ecological barrier* to LGT. A network of shared transposases among 774 microbial genomes supplies further support for the rarity of interhabitat gene transfers (Hooper et al. 2009). Halary et al. (2010) reconstructed a network of shared protein families among various genetic entities including microbial chromosomes, plasmids, and phage genomes. A comparison of network properties between plasmids and phage genomes revealed that plasmids are more frequently connected within the network in comparison to phages. From this they concluded that conjugation is more frequent than transduction in nature (Halary et al. 2010).

Trends and barriers to lexical borrowing

As in biology, there are certain barriers for horizontal transfer during language evolution. Since the sound systems of languages can be very different, the pronunciation difficulty of borrowed word within the recipient population may vary. In cases where the difference in the sound systems of donor and recipient languages is sufficiently large, direct borrowing will occure less frequently, resulting in a *similarity barrier*. For example, although the daily life in China is heavily influenced by Western culture, English words for Western concepts are rarely directly transferred into the Chinese language, but rather reproduced due to the large differences in the sound systems of English and Chinese. This is probably the reason why the borrowing capacity of Mandarin Chinese is the lowest in the sample of WOLD with only 1.2 % direct borrowings out of a total of 2042 words (Wiebusch 2009): Although in Standard Chinese there are many terms which have been coined under the influence of foreign examples, these words are expressed in a seemingly genuine manner. For example, the Standard Chinese word for "boomerang" is *fēiqùláiqì*, which literally can be translated as "there-and-back-flying device".

Borrowing events will be less frequent between geographically distant speech communities, resulting in a *spatial barrier*. The spatial barrier is closely connected with what one might call a *socio-cultural* or *socio-political barrier* for lexical borrowing: Due to social, cultural, or political reasons a given language variety may either be promoted or marginalized by the ones who speak it, resulting in a high or low borrowing rates (Tadmor 2009).

Furthermore, given that most borrowed words are due to the lack of certain words for certain concepts in the recipient language, that are present in the donor language, borrowing heavily depends on the meaning of the items being borrowed. While the exchange of innovations between different communities is often also accompanied by the exchange of lexical items, words denoting basic concepts that are essential for human life are less likely to be exchanged, resulting in a *functional barrier* (Hock and Joseph 2009).

OUTLOOK

Given the fact that language history and genome evolution take place in very different domains, it is not surprising to find many differences between both processes, especially when dealing with the details of the mechanisms and their explanations. From a more abstract perspective, there are, however, many interesting similarities between language history and genome evolution that are revealed when comparing the trends and the barriers to lateral transfer in the linguistic and the biological domains.

Understanding how languages and genomes change, how they eventually split, separate, and diverge, is a challenging problem in evolutionary biology and historical linguistics. Many different methods have been proposed so far to study these processes in detail. During the last two decades, linguists have especially focused on quantifying the traditional qualitative methods. As a result, many new approaches to language phylogenetic reconstruction have been proposed, leading to a better understanding of the genealogical processes that led to the diversification of different language families. However, because language change is not only based on the modification of inherited items but is also driven by the direct or indirect transfer of linguistic units, phylogenetic trees do not tell us the true story about the history of languages, but provide only a reduced version which may be often misleading. The same holds for microbial evolution where the high frequency of LGT renders the tree model insuficient. In both disciplines network approaches can assist to uncover on reticulated evolutionary events that were previously ignored.

REFERENCES

Babic, A.; Berkmen, M.B.; Lee, C.A. & A.D. Grossman (2011) 'Efficient Gene Transfer in Bacterial Cell Chains', mBio 2, e00027–11–e00027–11.

Babic, A.; Lindner, A.B.; Vulic, M.; Stewart, E.J. & M. Radman (2008) 'Direct Visualization of Horizontal Gene Transfer', Science 319: 1533–6.

Bapteste, E.; O'Malley, M. & R. Beiko (2009) 'Prokaryotic evolution and the tree of life are two different things', Biol Direct PMID: 19788731.

Beiko, R.G.; Harlow, T.J. & M.A. Ragan (2005) 'Highways of gene sharing in prokaryotes', Proc Natl Acad Sci USA 102: 14332–7.

Berglund, E.C.; Frank, A.C.; Calteau, A.; Vinnere Pettersson, O.; Granberg, F.; Eriksson, A.-S.; Näslund, K.; Holmberg, M.; Lindroos, H. & S.G.E. Andersson (2009) 'Run-Off Replication of Host-Adaptability Genes Is Associated with Gene Transfer Agents in the Genome of Mouse-Infecting Bartonella grahamii', PLoS Genet 5, e1000546.

Bonfante, G. (1931) 'I dialetti indoeuropei', in Annali del R. Instituto Orientale di Napoli 4: 69–185.

Burke GR, Moran NA (2011) 'Massive genomic decay in Serratia symbiotica, a recently evolved symbiont of aphids', Genome Biol Evol 3: 195–208.

Bryant, D.; Filimon, F. & R.D. Gray (2005) 'Untangling our past: Languages, Trees, Splits and Networks', in R. Mace, C.J. Holden & S. Shennan (eds), The evolution of cultural diversity: A phylogenetic approach (London: UCL Press): 69–85.

Cannon, W.F. (1960) 'The uniformitarian-catastrophist debate', Isis 51(1): 38–55.

Chan, C.X.; Beiko, R.G. & M.A. Ragan (2011) 'Lateral Transfer of Genes and Gene Fragments in Staphylococcus Extends beyond Mobile Elements', Journal of Bacteriology 193: 3964–77.

Chen I, & D. Dubnau (2004) 'DNA uptake during bacterial transformation', Nat Rev Microbiol 2: 241–9.

Chen, I.; Christie, P. & D. Dubnau (2005) The ins and outs of DNA transfer in bacteria. Science 2(310/5753): 1456–60.

Coscolla M.; Comas I. & F. Gonza lez-Candelas (2011) 'Quantifying nonvertical inheritance in the evolution of Legionella pneumophila', Mol Biol Evol 28: 985–1001.

Dagan, T. (2011) 'Phylogenomic networks', Trends in Microbiology 19: 483–91.

Dagan, T. & W. Martin (2007) 'Ancestral genome sizes specify the minimum rate of lateral gene transfer during prokaryote evolution', Proc Natl Acad Sci USA 104: 870–5.

Dagan, T.; Artzy-Randrup, Y. & W. Martin (2008) 'Modular networks and cumulative impact of lateral transfer in prokaryote genome evolution', Proc Natl Acad Sci USA 10039–44.

Darwin, C. (1859) On the origin of species by means of natural selection, or, the preservation of favoured races in the struggle for life (London: John Murray).

Dickerson, R. (1980) 'Evolution and gene transfer in purple photosynthetic bacteria', Nature 283(5743): 210–2.

Doolittle, W. Ford (1999) 'Phylogenetic classification and the universal tree', Science 284(5423): 2124–9.

Dubey, G.P. & S. Ben-Yehuda (2011) 'Intercellular Nanotubes Mediate Bacterial Communication', Cell 144: 590–600.

Dyen, I.; Kruskal, J.B. & P. Black (1992) 'An Indoeuropean classification: A lexicostatistical experiment', Trans. Am. Philos. Soc. 82: 1–132.

Freeman, V.J. (1951) 'Studies on the virulence of bacteriophage-infected strains of Corynebacterium diphtheriae', Journal of Bacteriology 61: 675.

Garcia-Vallvé, S.; Romeu, A. & J. Palau (2000) 'Horizontal gene transfer in bacterial and archaeal complete genomes', Genome Research 10: 1719–25.

Gèvaudan, P. (2007) Typologie des lexikalischen Wandels. Bedeutungswandel, Wortbildung und Entlehnung am Beispiel der romanischen Sprachen (Tübingen: Stauffenburg).

Gogarten J.P.; Doolittle, W.F. & J.G. Lawrence (2002) 'Prokaryotic evolution in light of gene transfer, Molecular Biology and Evolution 19(12): 2226–38.

Halary, S.; Leigh, J.W.; Cheaib, B.; Lopez, P. & E. Bapteste (2010) 'Network analyses structure genetic diversity in independent genetic worlds', Proc Natl Acad Sci USA 107: 127–32.

Hao, W. & G.B. Golding (2006) 'The fate of laterally transferred genes: life in the fast lane to adaptation or death', Genome Res, 16: 636–643.

Hamed, M.B. & F. Wang (2006) 'Stuck in the forest: Trees, networks and Chinese dialects', Diachronica 23: 29–60(32).

Haspelmath, M. & U. Tadmor (eds) (2009) Loanwords in the World`s Languages. A Comparative Handbook (Berlin/New York: de Gruyter).

Hirt, H. (1905) Die Indogermanen. Ihre Verbreitung, ihre Urheimat und ihre Kultur, Vol. 1 (Strassburg: Trübner).

Hock, H.H. & B.D. Joseph (1995 [2009]) Language history, language change and language relationship. An introduction to historical and comparative linguistics, 2nd ed. (Berlin/New York: de Gruyter).

Holden, C.J. & S. Shennan (2005) The evolution of cultural diversity (London: UCL Press): 67–84.

Huson, D.H. & C. Scornavacca (2011) 'A Survey of Combinatorial Methods for Phylogenetic Networks', Genome Biology and Evolution 3: 23–35.

Jain, R.; Rivera, M.C. & J.A. Lake (1999) 'Horizontal gene transfer among genomes: the complexity hypothesis', Proc Natl Acad Sci USA 96: 3801–6.

Kenzaka, T., Tani, K. & M. Nasu (2010) 'High-frequency phage-mediated gene transfer in freshwater environments determined at single-cell level', The ISME Journal 4: 648–59.

Kloesges, T., O. Popa, W. Martin & T. Dagan (2011) 'Networks of gene sharing among 329 proteobacterial genomes reveal differences in lateral gene transfer frequency at different phylogenetic depths', Molecular Biology and Evolution 28: 1057–74.

Koonin, E.V. (2009) 'Darwinian evolution in the light of genomics', Nucleic Acids Research 37: 1011–34.

Kunin, V. (2005) 'The net of life: Reconstructing the microbial phylogenetic network', Genome Research 15: 954–9.

Lang, A.S. & J.T. Beatty (2007) 'Importance of widespread gene transfer agent genes in alphaproteobacteria', Trends in Microbiology 15: 54–62.

Lang, A.S.; Zhaxybayeva, O. & J.T. Beatty (2012) 'Gene transfer agents: phage-like elements of genetic exchange', Nat Rev Micro 10: 472–82.

Lawrence, J.G. & H. Ochman (1998) 'Molecular archaeology of the Escherichia coli genome', Proc Natl Acad Sci USA 95: 9413–7.

Lima-Mendez, G., J. Van Helden, A. Toussaint & R. Leplae (2008) 'Reticulate Representation of Evolutionary and Functional Relationships between Phage Genomes', Molecular Biology and Evolution 25: 762–77.

Majewski, J. & F.M. Cohan (1999) 'DNA sequence similarity requirements for interspecific recombination in Bacillus', Genetics 153: 1525–33.

Martin, W. (1999) 'Mosaic bacterial chromosomes: a challenge en route to a tree of genomes', Bioessays 21: 99–104.

McDaniel, L.D.; Young, E.; Delaney, J.; Ruhnau, F.; Ritchie, K.B. & J.H. Paul (2010) 'High Frequency of Horizontal Gene Transfer in the Oceans', Science 330: 50.

McMahon, A. & R. McMahon (2005) Language classification by numbers (Oxford: Oxford University Press).

Mirkin, B.G.; Fenner, T.I.; Galperin, M.Y. & E.V. Koonin (2003) 'Algorithms for computing parsimonious evolutionary scenarios for genome evolution, the last universal common ancestor and dominance of horizontal gene transfer in the evolution of prokaryotes', BMC Evol Biol 3: 2.

Moran, N.A.; McLaughlin, H.J. & R. Sorek (2009) 'The dynamics and time scale of ongoing genomic erosion in symbiotic bacteria', Science 323: 379–82.

Nakamura, Y.; Itoh, T.; Matsuda, H. & T. Gojobori (2004) 'Biased biological functions of horizontally transferred genes in prokaryotic genomes', Nat Genet 36: 760–6.

Nakhleh, L.; Ringe, D. & T. Warnow (2005) 'Perfect Phylogenetic Networks: A New Methodology for Reconstructing the Evolutionary History of Natural Languages', Language 81.2: 382–420.

Navarre, W.W.; Porwollik, S.; Wang, Y.; McClelland, M.; Rosen, H.; Libby, S.J. & F.C. Fang (2006) 'Selective silencing of foreign DNA with low GC content by the H-NS protein in Salmonella', Science 313: 236–8.

Nelson-Sathi, S.; List, J.-M.; Geisler, H.; Fangerau, H.; Gray, R.D.; Martin, W. & T. Dagan (2011) 'Networks uncover hidden lexical borrowing in Indo-European language evolution', Proceedings of the Royal Society B. Biological Sciences 278.1713: 1794–803.

Norman, A., Hansen, L.H. & S.J. Sørensen (2009) 'Conjugative plasmids: vessels of the communal gene pool', Philosophical Transactions of the Royal Society B, Biological Sciences 364: 2275–89.

Ochman, H.; Lawrence, J.G. & E.A. Groisman (2000) 'Lateral gene transfer and the nature of bacterial innovation', Nature 405: 299–304.

Pagel, M. (2009) 'Human language as a culturally transmitted replicator', Nature Reviews Genetics 10: 405–15.

Pal, C; Papp, B. & M.J. Lercher (2005) 'Adaptive evolution of bacterial metabolic networks by horizontal gene transfer', Nat Genet 37: 1372–5.

Popa, O. & T. Dagan (2011) 'Trends and barriers to lateral gene transfer in prokaryotes', Current Opinion in Microbiology 14(5): 615–23

Popa, O.; Hazkani-Covo, E.; Landan, G.; Martin, W. & T. Dagan (2011) 'Directed networks reveal genomic barriers and DNA repair bypasses to lateral gene transfer among prokaryotes', Genome Research 21: 599–609.

Puigbo, P.; Wolf, Y.I. & E.V. Koonin (2010) 'The Tree and Net Components of Prokaryote Evolution', Genome Biology and Evolution 2: 745–56.

Ragan, M.A. (2009) 'Trees and networks before and after Darwin', Biol Direct 4: 43.

Schleicher, A. (1853a) 'Die ersten Spaltungen des indogermanischen Urvolkes', Allgemeine Monatsschrift für Wissenschaft und Literatur Sept 1853: 786–87.

Schleicher, A. (1853b) 'O jazyku litevském, zvláště na slovanský. Čteno v posezení sekcí filologické král', České Společnosti Nauk dne 6. června 1853, Časopis Čsekého Museum 27: 320–34.

Schleicher, A. (1863) Die Darwinsche Theorie und die Sprachwissenschaft. Offenes Sendschreiben an Herrn Dr. Ernst Haeckel (Weimar: Böhlau).

Schmidt, J. (1872) Die Verwantschaftsverhältnisse der indogermanischen Sprachen (Weimar: Böhlau).

Schuchardt, H. (1900[1870]) Über die Klassifikation der romanischen Mundarten, Probe-Vorlesung gehalten zu Leipzig am 30. April 1870 (Graz: Styria).

Smith, H.O.; Tomb, J.F.; Dougherty, B.A.; Fleischmann R.D. & J.C. Venter (1995) 'Frequency and distribution of DNA uptake signal sequences in the Haemophilus influenzae Rd genome', Science 269: 538–540

Sorek, R.; Zhu, Y.; Creevey, C.; Francino, M. & P. Bork (2007) 'Genome-wide experimental determination of barriers to horizontal gene transfer', Science 318: 1449–52.

Southworth, F.C. (1964) 'Family-tree diagrams', Language 40.4: 557–65.

Sullivan, M.B.; Lindell, D.; Lee, J.A.; Thompson, L.R.; Bielawski, J.P. & S.W. Chisholm (2006) 'Prevalence and Evolution of Core Photosystem II Genes in Marine Cyanobacterial Viruses and Their Hosts', PLoS Biol 4(8): e234.

Sutrop, U. (1999) 'Diskussionsbeiträge zur Stammbaumtheorie', Fenno-Ugristica 22: 223–51.

Tadmor, U. (2009) 'Loanwords in the world's languages. Findings and results', in M. Haspelmath & U. Tadmor (eds), Loanwords in the world's languages. A comparative handbook (Berlin/New York: de Gruyter): 55–75.

Thomas, C.M. & K.M. Nielsen (2005) 'Mechanisms of, and Barriers to, Horizontal Gene Transfer between Bacteria', Nat Rev Micro 3: 711–21.

Weinreich, U. (1953 [1974]) Languages in contact, With a preface by André Martinet, 8th ed (The Hague / Paris: Mouton).

Wells, R.S. (1973) 'Lexicostatistics in the regency period', in Lexicostatistics in Genetic Linguistics, Proceedings of the Yale Conference, 3./4. April 1971 (Yale University): 118–21.

Wiebusch, Thekla (2009) 'Mandarin Chinese vocabulary', M. Haspelmath & U. Tadmor (eds), World Loanword Database (Munich: Max Planck Digital Library). URL: http://wold.living-sources.org.

Zhao, Y., Wang, K., Ackermann, H.-W., Halden, R.U., Jiao, N., and Chen, F. (2010). Searching for a "Hidden" Prophage in a Marine Bacterium. Applied and Environmental Microbiology 76, 589–595.

Zhaxybayeva, O.; Gogarten, J.P.; Charlebois, R.L.; Doolittle, W.F. & R.T. Papke (2006) 'Phylogenetic analyses of cyanobacterial genomes: quantification of horizontal gene transfer events', Genome Res 16: 1099–1108.

Zhaxybayeva, O. & W.F. Doolittle (2011) 'Lateral gene transfer', Curr Biol 21: R242-R246.

A PRELIMINARY CASE FOR EXPLORATORY NETWORKS IN BIOLOGY AND LINGUISTICS: THE PHONETIC NETWORK OF CHINESE WORDS AS A CASE-STUDY

Philippe Lopez, Johann-Mattis List, Eric Bapteste

THE RAISE OF NETWORKS AS COMPARATIVE METHODS IN EVOLUTIONARY BIOLOGY

Linguists, as well as biologists, study historical objects that form lineages, undergoing transformations over time. Biologists, as well as linguists, therefore, are very dependent on comparative analyses to structure and analyze their data. Thus, it seems intuitive that conceptual and methodological researches in both fields could inform each other, and benefit to both fields. In particular, the comparative approaches elaborated in biology are experiencing massive developments that could be explored in linguistic studies.

In biology, the general picture of evolution is becoming increasingly complex. Evolutionary innovations and changes are effected both by processes of vertical descent and introgressive (or combinatory) processes (recombination, lateral gene transfer, symbioses) (Bapteste et al. 2012; Bapteste et al. 2009; Dagan et al. 2008; Dagan and Martin 2009; Huang and Gogarten 2007; Kloesges et al. 2011; Marin et al. 2005; Wu et al. 2011). Vertical descent processes are usually modeled and studied using a common tree (e.g. a gene or a species tree) (O'Malley 2011). By contrast, combinatory processes reassort, regroup or merge evolutionary objects. Examples include mosaic genes, genomes and intricate symbiotic associations, and coalitions based on multiple lineages, persisting via the tight co-evolution of evolutionary players from distinct lineages (Bapteste et al. 2012) (e.g. cells and mobile genetic elements such as plasmids and phages in multispecies biofilms (Ghigo 2001; Hall-Stoodley et al. 2004; Periasamy 2009; Wintermute 2010) or in the gut microbiomes (Jones 2010; Lozupone et al. 2008; Qu et al. 2008; Martin et al. 2007)). Thus original genetic associations from multiple sources, sustained by a diversity of evolutionary processes, can be cemented into novel evolutionary units, i.e. when the transfer of domains produces new genes, or the transfer of genes produces new gene clusters, pathways and mosaic genomes. Likewise genetic associations between distantly related entities can evolve into novel symbiotic organisms and microbial coalitions (Dagan et al. 2008; Martin et al. 2007; Moustafa et al. 2009).

Consequently, the usual framework of a single tree fails to represent the evolution of many biological entities, at different biological scales, in particular when these entities are mergers from multiple lineages. Problems also arise when the investigated entities are too divergent from reference entities already put in a reference tree to be simultaneously analyzed on the same tree. Highly divergent entities will typically not be readily comparable with reference entities in a single analysis, as the homology between divergent entities becomes too distant to be effectively detected. This problem does not constitute a major limit for network analyses, which can handle higher levels of divergence (Alvarez-Ponce et al. 2013; Beauregard-Racin et al. 2011). Thus, in biology networks are increasingly used as alternative models to describe more of the complexity of biological evolution (Bapteste et al. 2012; Dagan et al. 2008; Dagan 2009; Alvarez-Ponce et al. 2013; Beauregard-Racine et al. 2011; Fondi and Fani 2010; Halary et al. 2010; Lima-Mendez et al. 2008); Skippington and Ragan 2011. These methods are an invaluable complement to the construction of common trees of single lineages of objects from a given level of organisation (e.g. gene trees focusing on genes, organismal trees focusing on organisms, species trees focusing on species, etc.). Moreover, their potential to provide a novel analytical framework for exploratory evolutionary studies is also increasingly acknowledged.

Indeed, networks are more flexible graphs than trees. They are less constrained in their representation of the data and the relationships between objects, and can support different levels of abstraction. Typically, a network G, noted G = (V, E), comprises a set V of vertices or nodes associated with a set E of edges. The nature of the nodes (e.g. a domain, a gene, a gene cluster, a genome, an environment) as well as their rules of connection can be used as parameters that vary in exploratory analyses (Burian 2011). Thus, using networks of genes, or of genomes, or of lineages, or of environments, biological diversity can be observed at many levels, e.g. within one (or many) gene families, genomes, lineages, communities, or environments (Zhaxybayeva and Doolittle 2011), by simply varying the nature of the investigated nodes. Moreover, each level of biological diversity can be structured in different informative ways by changing the types of edges represented in these graphs. For instance, for nodes corresponding to the same set of protein sequences, a graph could either only show connections retracing functional interactions between these proteins (Martha et al. 2011; Vinayagam et al. 2011; Wang et al. 2011), or connections reflecting only genealogical relationships between these proteins (Beauregard-Racine et al. 2011; Alvarez-Ponce and McInernex 2011), etc. When these variations in the type of edges represented in these networks induce changes in the graph topology between the nodes, networks comparisons can identify robust/transient patterns of connections, appearing over a large/limited range of conditions / biological levels, i.e. transient functional interactions between unrelated proteins.

Interestingly, these remarkable patterns need not necessarily be a priori expected. Network-based studies of genetic diversity typically foster the discovery of many unrecognized patterns, and thus contribute to actively generate novel hypotheses about the evolution of genetic diversity. For example, in a gene network (Beauregard-Racine et al. 2011; Bittner et al. 2010), nodes are gene sequences, connected by weighted edges when they share a relationship of homology/identity, as assessed by a BLAST score. Each gene family is easily characterized as it falls in a separated connected component. When environmental sequences are included in such analyses along with sequences from cultured organisms, novel environmental gene families can be discovered. Moreover, such gene networks can be used to detect evolutionary units that a tree of sequences alone cannot detect. For instance, the study of 10^4–10^6 sequences allows to detect groups of genes families with complex evolutionary patterns (expansions, high evolutionary rates, combinations (Beauregard-Racine et al. 2011), etc.), fused genes (Gallagher and Eliassi-Rad 2008; Jachiet et al. 2013), and very distantly related gene forms (Alvarez-Ponce et al. 2013), branching off rather than within the sequences from known gene families. Hence, an exploratory approach of genetic diversity can unravel unsuspected highly divergent gene forms, and questions what their biological function might be.

The exploratory use of network can be formalized in even more general terms. While most evolutionary studies are mainly concerned by the justification of theories about how entities diverge or by the test of genealogical hypotheses seeking to establish sister-group relationships, exploratory sciences try to develop new concepts to 'fix any evolutionary phenomenon' calling for explanation (Burian 2011; Franklin-Hall 2005). It uses networks to establish and classify relevant patterns that had not yet been well characterized, such as the patterns that a tree-based approach would fail to represent. Networks can not only quickly sort massive amount of data with limited *a priori* on the connections between the objects analyzed in these data, but also rapidly expose their potential underlying (intriguing) patterns/structures.

Finally, networks offer a precious mathematical framework for comparative and exploratory analyses, because the topological properties of their nodes and edges (Koschützki 2008) can be computed and compared. Topological indices, such as the conductance (Leskovec et al. 2008) of a group of nodes can be estimated. For a given group of nodes (e.g. nodes corresponding to words from a given cognate, or to genes from a given gene family), the conductance C is computed as: $C = N_{ext} / (N_{ext} + 2*N_{int})$, where N_{int} is the number of internal edges (e.g. linking members of that cognate or gene family) and N_{ext} is the number of external edges (e.g. linking a member from that cognate/gene family and a member from another cognate/gene family). Clustered and/or isolated groups (e.g. of words from the same cognate or of genes from the same gene family) have a conductance close to 0, while spread out or fragmented groups have a conductance close to 1. Thus, the conductance measures whether nodes with a given label cluster in the networks (i.e. whether words from the same cognate, grammatical class or dialect are more similar to one another than to any other words; or whether genes with the same function, or from the same genus, or from the same environment, are more similar to one another than to any other gene).

Importantly, the increasingly recognized diversity of biological evolutionary processes and patterns observed in biological studies may also find some echo in the field of linguistics. This latter discipline also inquires history and evolution of numerous evolving entities, such as word families and languages, which may very well be effected by vertical and combinatory processes (Nelson-Sathi et al. 2011) (see Table 1, for a possible analogy between the evolution of biological and linguistics objects). Therefore, we wanted to test here whether the study of some linguistic objects using networks could foster novel hypotheses about their evolution, and offer a test-case for the relevance of some of the analogies between biological and linguistic objects. More precisely, we used a dataset of 48 semantic glosses translated into 40 Chinese dialects to reconstruct a word network based on phonetic similarity. We classified these words by their meaning, dialect of origin and grammatical categories, and estimated the conductance (e.g. the phonetic consistency) of each meaning, dialect, and grammatical category. We observed that different selective (sociological and linguistic) pressures are acting on how a word sounds, introducing phonetic variability and structure in Chinese languages according to different rules, influenced by the grammatical category to which the words belong. Yet, cognate sets and gene families present rather different levels of diversity (phonetic and genetic, respectively), encouraging the innovative development of specific network methods in linguistics rather than the simple import of comparative methods of evolutionary biology that are currently better suited for biological objects.

Table 1. Some possible correspondence for an analogy between evolutionary biology and linguistics

Evolutionary Biology	Evolutionary Linguistics
Gene (particular function)	Word (meaning)
Gene family	Cognate set
Gene functional ontology	Grammatical category
Genome	Dialect, Language
Lateral Gene Transfer	Lexical borrowing
Genetic diversity	Phonetic diversity
Distant homology	Hidden cognacy
Selective pressures	Sociological, linguistic constraints

THEORETICAL POWER OF EXPLORATORY NETWORKS IN LINGUISTICS

Network-based analyses allow relaxing some *a priori* constraints generally imposed by tree-based analyses. Although disquieting in the first place for practitioners more trained to work within a tree-based framework, this reduction of constraints in data display offers a novel way to capture more of the evolutionary processes and patterns in addition to the process and pattern of 'vertical descent with modification'. This general observation, we believe, probably holds true for both evolutionary biology and linguistics, assuming that in both fields several processes cannot be properly represented and modeled with a tree-based approach, which inexorably constrains the analyses to be only expressed in terms of divergence and dichotomies, as well as the type of data suited for an evolutionary analysis. In molecular phylogenetics for instance, the suitable material are homologous sequences that align well with one another, since they belong to a single sequence family derived from an ancestral gene copy. This practice considerably restricts the amount of molecular data amenable for analysis, and thus the scope of the analysis (Bapteste et al. 2012; Bapteste et al. 2008; Dagan and Martin 2006; Leigh et al. 2011). Sequences undergoing more complex evolutionary processes are not included in the analyses, because they would blur the reconstruction of the gene (or species) genealogy that tree-based analysis generally aims for. Gene networks overcome this issue of massive *a priori* data exclusion, by allowing the display of more processes and relations between gene forms (although these are not only the usual relations of homology) than is permitted by a tree. Similarly, we argue that word networks may extend the scope of linguistic analysis beyond inferences focused on predefined cognate sets by recovering more distant cases of cognacy or by introducing novel measures of similarity (here phonetic distances) between words.

Various types of distances could be used to reconstruct a word network. In the present analysis, we focus on phonetic word networks, as a mean to display (and then later to analyze) the phonetic diversity of words within several dialects (Figure 1).

Figure 1: Virtual weighted cognate networks.

A. A component corresponding to a set of words that may, or may not, belong to an accepted cognate family. Nodes are words, color-coded based on their dialect of use. Edges are weighted according to any distance metric, i.e. a phonetic distance between pairs of words. **B-D.** Virtual component topologies that would support distinct interpretations on words evolution. B. A phonetically conserved word family, the typical pattern in a word network for a bona fide cognate set. C: The bridging node is an emerging word resulting from borrowing and fusion events of words from distinct dialects. D: A family of words with two strongly connected communities and peripheral nodes (light green, grey, and black), indicating distantly related versions of these words in several dialects, suggesting different evolutionary rates (of the sounds) of these words in the dialects, while showing some 'regional' conservation in the brown dialects.

Such word networks are disconnected, because each set of words using a common pool of phonemes will create its own connected component in the graph. Indeed, when two words are phonetically different (e.g. presenting less than a minimal phonetic similarity with one another), no edge is drawn between them, and they fall into distinct subgraphs within the word network. Otherwise, when two words display some phonetic similarity (e.g. when their phonetic distance is lower than a given threshold), these words are connected by weighted edges, with an edge weight that is inversely proportional to the phonetic distance, so that words closer in phonetic distance have edges with higher weights on the graph.

A notable consequence of the great inclusiveness of such phonetic word networks is that not all their components have to be cliques, i.e. maximally connected components in which each and every node directly connects to each and every other node in the "word family". In biological networks (gene networks), the clique pattern is typical for ubiquitous and conserved gene families, in which all sequences are highly similar because inherited from a last common ancestor and affected by a relatively limited amount of mutations so that their homology can still be successfully assessed. However, it is an empirical question whether, in word networks, the analogs of gene families, the cognates, will also produce cliques, with related words sounding sufficiently alike to be all directly connected together to the exclusion of other unrelated words. More or less structured connections can emerge in these graphs showing phonetic distances between words. In particular, cognates may not produce cliques, and belong to components that are either the result of more complex evolutionary processes than vertical descent from an ancestral word alone (e.g. the evolution of these words and word families may involve combinatory processes); or they could belong to components joining groups of words affected by phonetic convergences (a phenomenon that is expected because a typical word is short and the phoneme diversity is limited). These latter components may mix together words belonging to different cognate sets, however connected because they exploit overlapping pools of "phonemes". Finally, members of the same cognate set may also be highly disconnected in the network, if those cognates are word families in which sounds evolve very fast, to a point that it becomes impossible to detect the common historical origin of these words based on phonetic distances alone. A further investigation of the classes of words (organized by dialects, grammar and meanings) may unravel some rules of "phoneme" associations and the constraints that may affect how words sound. Here, we investigated phonetic diversity from multiple perspectives to test whether and how the dialect of origin of a word, or its grammatical function, or its meaning affected its phonetic consistency.

APPLICATION TO THE NETWORK OF CHINESE COGNATES

We used a subpart of Hóu's collection of Chinese dialect data (Hóu 2004), consisting of 48 semantic glosses translated into 40 Chinese dialects. The whole data comprised 2,999 different words. Following cognate judgments provided in the original data, these words were grouped into 337 different cognate sets. We further calculated phonetic distances between all words using the SCA method (List 2012) to derive alignment scores and the formula by Downey et al. (Downey et al. 2008) to convert similarity into distance scores. Mean distance between any two words was estimated to be 1.17, but only 0.35 between two cognates (Figure 2).

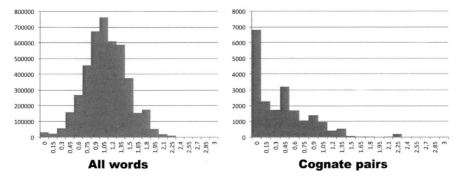

Figure 2: Phonetic distance between pairs of words estimated by SCA.

Left panel: Phonetic distance between all pairs of words; Right panel: Phonetic distance between pairs of cognates. This graph shows that words belonging to the same cognate set sound much more alike than random pairs of words, however they can still show important phonetic variations.

This simple observation indicated that words belonging to the same cognate set sound much more alike than random pairs of words, a property that is also observed in evolutionary biology. Extant sequences diverging from the same ancestral sequence (homologous sequences) are expected to be much more similar to each other than to any unrelated sequence and given the extremely low probability that two random sequences will show some similarity by chance alone, molecular evolutionists usually consider that analogy implies homology. We thus investigated whether similarity networks applied to linguistic data would perform in a similar way as they do in molecular evolution. Just as a threshold is needed to determine whether two sequences should be considered homologous, a maximum phonetic distance has to be used to determine if two words show significant phonetic proximity. Since the average distance between words from the same cognate was 0.35, we used that threshold to build our network and linked pair of words that showed a phonetic distance lower or equal to 0.35. This protocol allowed us to filter the 4.5 million potential edges (for 2,999 words) of the most inclusive word network to reduce it to its most pronounced relationships, summarized by about 60,000 edges encompassing 97 % of the word dataset. This phonetic diversity was then refined by distinguishing two kinds of edges in the reduced network: on the one hand, cognate edges, connecting two members of the same cognate set, and on the other hand, similarity edges connecting members from different cognate sets. First, for representation purposes, we used only cognate edges (Figure 3). The resulting sub-network (i) very neatly split some cognates into different graph components (indicating that groups of words from the same cognate set can sound very differently) and (ii) showed components that were not cliques, demonstrating the importance of phonetic variation within cognate sets.

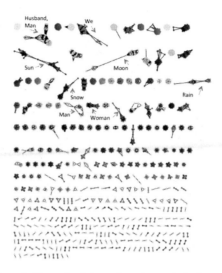

Figure 3: Network of close phonetic proximity between
words belonging to the same cognate set.

Nodes correspond to words, colored by meanings, connected by edges indicating a close phonetic proximity (distance < 0.35) between pairs of words from the same cognate set. Some meanings are indicated along this subnetwork. Some connected components are not cliques, indicating strong divergence.

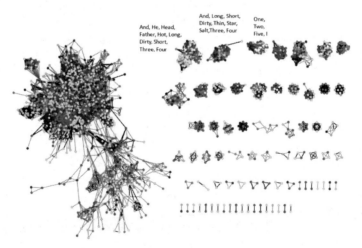

Figure 4: Network of close phonetic proximity with both cognate and similarity edges.

Nodes correspond to words, colored by meanings, connected by cognate edges (in black) and similarity edges (in grey) indicating a close phonetic proximity (distance < 0.35) between pairs of words. Colors are extremely scrambled, showing that the phonetic consistency (clustering and isolation) of most Chinese meanings is low.

However, one should not be mistaken by the exclusive focus on cognate edges, because when all phonetic comparisons (hence when both cognate and similarity edges) are considered, it is equally obvious that words from different cognate sets also sound very close (Figure 4). In other words, there are strong phonetic convergences between historically unrelated words. This observation brings forward a fundamental difference between cognate sets and their presumed analogs in the biological sciences: gene families. While gene families can be identified based on their genetic distances, cognate sets cannot be successfully identified based on their phonetic distances. In linguistics, this is known as the problem of *phenotypic* as opposed to *genotypic* similarity (Lass 1997). Phenotypic similarity refers to word similarities based on language-independent surface criteria by which the similarity of phonetic segments is determined. Genotypic similarity refers to similarity that is language-specific. That means that it can be only defined for distinct pairs of languages that are known to be genetically related. For a given language pair, genotypic similarity is determined in form of sound correspondences, that is sounds (phonemes) that are known to be homologous. As an example for such correspondences, compare the cognate words English *token* [təʊkən] and German *Zeichen* [tsaɪçən] 'sign'. Although these words sound very different, it is easy to show that the sounds regularly correspond to each other, as can be seen from English *weak* [wiːk] vs. German *weich* [vaɪç] 'soft' for the correspondence of [k] with [ç], and English *tongue* [tʌŋ] vs. German *Zunge* [tsʊŋə] 'tongue' for the correspondence of [t] with [ts]. Genotypic similarity is quite similar to the relation between a source text and its encryption, where all characters may refer regarding their substance, although they are related by an underlying distinct mapping.

Thus, while the alignment problem in biology can be stated under the assumption that two sequences are both drawn from the same alphabet (e.g. proteins), the alignment problem in linguistics is essentially the problem of aligning two sequences drawn from two *different* alphabets. Although from a general perspective cognate sets and gene families are the same kind of classes of objects, practically they cannot be detected, hence studied alike. Indeed, members of both cognate sets and gene families share the extrinsic, relational, property of originating from the same common ancestor; yet this historical essence of cognates and of gene families does not translate into the definition of sets of objects with intrinsic exclusive properties. There is no obligate (nor strong) correlation between 'having the same origin' and 'sounding alike' for words, while there is a stronger correlation between 'having the same origin' and 'having closer genetic sequences than anyone else' for genes. This difference implies that while the classification of genes into gene families allows for some generalizations about the members of the gene family, the classification of words into cognate sets allows for less generalization regarding their phonetic similarity. This difference may not come as a surprise, since the limited amount of phonemes and the limited size of words (as opposed to the high combinatorials of DNA bases in relatively longer sequences) makes such convergences expected. However, this high level of phonetic convergence means that word network based on phonetic distances are unlikely to be as discriminating as gene networks based on genetic distances. While the latter can be used to infer classes of undetected (hidden) homology, it seems more problematic to use the former to infer undetected (hidden) cognacy, without the development of specific, genotypic, distances, adapted to the objects of linguistics. Moreover, the fact that gene families can be defined both based on correlated extrinsic and intrinsic properties, while cognate sets seem to be mostly characterized by extrinsic properties raises questions on whether these objects can be used alike for explanations, descriptions and inductive inferences in the fields of biology and linguistics, and if not, whether the analogy between cognate and gene families should not be also refined.

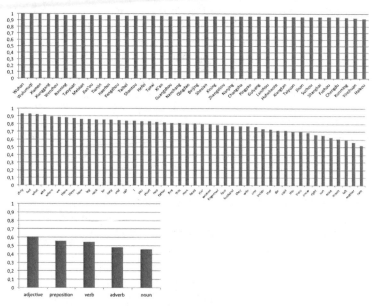

Figure 5: Conductance of the meanings, dialects and grammatical types in the Chinese word network.

Conductance for each item (x-axis) is indicated on the y-axis, and computed as described in the text. Significance was assessed by shuffling the labels of the original network, then computing the various conductances (Top: dialect, Middle: meaning, Bottom: grammar) on this randomized network. The procedure was repeated 1,000 times to obtain a normal distribution of conductances for random classes of the same size as the tested classes. Except for the 11 leftmost dialects (Wuhan to Haerbin), all observed conductances were less than 2 sigma below the mean of their corresponding normal distribution, meaning that the estimated conductance was not a mere effect of the sample size.

Importantly however, the fact that the Chinese word network based on phonetic distances is not strongly structured by cognate sets does not mean that this network does not show another type of informative structure. We classified the words into three functional categories to test whether, in spite of this high amount of phonetic convergence, phonetic properties of the words were not random, suggesting some rules and selective pressures on phoneme combinations. To this end, each word was labeled based on its meaning, dialect of origin, and grammatical type (Adjective,

Adverb, Noun, Preposition, Verb), and the conductance of each of these labels in the network was assessed (Figure 5). All dialects presented a very high conductance (close to 1) indicating that they cannot easily be distinguished based on the phonetic similarity between their words: words from different dialects can sound very close, or to put it differently, no dialect shows strong phonetic consistency. This is not surprising, since high phonetic similarity between the words of a single language would make it difficult for the speakers to communicate. Meanings also had high conductances, however lower than those of dialects, and some meanings ("right", "sun", "nose", "moon", "left", "mother" and "rain") had even relatively low conductances, testifying that some combinations of phonemes were preferentially associated with these meanings. Given that words denoting these meanings are usually highly preserved, often going back to the same ancestor form in all Chinese dialects, this result is also not unexpected. Strikingly, grammatical types present the lowest conductances of all the tested classes and structure the similarity network more strongly than all the other types we checked (Figure 6). A mechanical explanation for this might lie in the distance measure that we used. Although normalized for word length, the distance measure is still rather sensitive to the comparison of words that have an equal length, yielding lower distance scores for words with a similar length. Since average word length tends to be very similar for parts of speech in Chinese (prepositions usually consist only of one syllable, nouns usually have two syllables), this may also be a reason for the low conductance of words corresponding to the same part of speech.

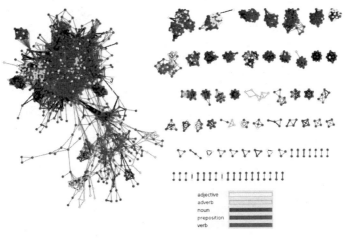

Figure 6: Network of close phonetic proximity with both cognate and similarity edges colored by grammatical types.

Nodes correspond to words, colored by grammatical types, connected by cognate edges (in black) and similarity edges (in grey) indicating a close phonetic proximity (distance < 0.35) between pairs of words. This figure can be contrasted to Figure 4 to verify that meanings are less phonetically structured than grammatical types.

CONCLUSION & PERSPECTIVES

In both the fields of evolutionary biology and linguistics, graphs appear as excellent tools for the exploratory analysis of evolving objects, be they words or genes. However, because the evolution of words is affected by more convergences than the evolution of genes, specific adjustments seem still to be required to integrate networks in the toolkit of linguistic studies. In particular, unsupervised automatic cognate detection might prove much harder than gene family detection. The investigation of the phonetic diversity of cognate words from Chinese dialects with relatively simple networks however was already powerful enough to identify different phonetic structures at different levels of linguistic organization. Dialects, meanings and grammatical categories seem subjected to distinct intensities of selective pressures, affecting the diversity of phonemes used in their making, in ways that now deserved to be explained.

MATERIAL AND METHODS

Dataset

The data that we used for our analysis is taken from the Hànyǔ Fāngyán Yīnkù (Hóu 2004), a CD-ROM that offers many different resources on Chinese dialects, including phonological descriptions, phonetic transcriptions, and sound recordings for 40 different dialect varieties. From the CD-ROM, we extracted a part of the lexical subset, consisting of 48 glosses ("concepts") translated into the 40 varieties. These 48 glosses belong to the basic vocabulary in the strict sense of Swadesh (Swadesh 1952; Swadesh 1955). Chinese dialects often have a lot of synonyms for the very same concept; therefore the resulting dataset is made of 2,999 words in total. The source material was given in a format not tractable for computational analyses. Therefore, the extraction procedure was carried out semi-automatically, applying additional manual cleaning. All entries were double-checked by comparing the phonetic transcription for each word with its corresponding sound recording. The data was further enriched by looking up the grammatical categories of the glosses, translating the glosses into English, adapting the phonetic transcriptions to plain IPA, and applying a rough procedure for automatic cognate detection that is described in the following section.

Cognate Judgments

In Chinese dialectology it is common to give not only the pronunciation of a given dialect word, but also an assessment regarding its homology. Homology assessments are usually coded by giving the Chinese characters corresponding to a given word. Since for most Chinese characters the Middle Chinese readings (spoken around the 6th century) can be reconstructed from old rhyme books, a character is somewhat similar to a proto-form. Thus, Táoyuán [ŋit²²tʰeu¹¹] and Hǎikǒu [zit³hau³¹] "sun" are both written as 日头, and the proto-form would have been pronounced as *n̩it⁴duw¹ in Middle Chinese times (if the compound was already present during that time). Note that the character assignments in Chinese dialectology are homologs in the strict sense, since no distinction between borrowing and vertical inheritance is drawn. Using this procedure, the 2,999 words could be grouped into 337 cognate sets.

Phonetic distances

Phonetic distances between all words were calculated using the SCA method (List 2012) to derive alignment scores, and the formula by Downey et al. (Downey et al. 2008) to convert similarity into distance scores. The resulting distance measure is "phenotypic" in the sense of Lass (Lass 1997) in so far as it is language-independent, neglecting the presence or absence of previously established sound-correspondence patterns. However, it is based on an enhanced function for the scoring of phonetic segments, and previous studies (List 2012) could show that it outperforms alternative distances measures, such as the normalized Levenshtein distance (Levenshtein et al. 2010), or the measure underlying the cognate detection method by Turchin et al. (Turchin et al. 2010). Therefore, this distance measure seems to be a more reliable basis for network applications than alternative ones.

Network visualization and analyses

The network layouts were produced by Cytoscape software (Smoot et al. 2011), using force directed layouts. Conductances were computed as: $C = N_{ext} / (N_{ext} + 2*N_{int})$, where N_{int} is the number of internal edges (e.g. between members of that cognate or gene family) and N_{ext} is the number of external edges (e.g. between a member from that cognate/gene family and a member from another cognate/gene family). Significance of these conductances was assessed by shuffling the labels of the original network, then computing the various conductances (dialect, meaning, grammar) on this randomized network. The procedure was repeated 1,000 times to obtain a normal distribution of conductances for random classes of the same size than the tested classes. Unless specified otherwise, most observed conductances were more than 2 sigma lower than the mean of their corresponding normal distribution, meaning that the conductance values are not a mere effect of the sample size.

REFERENCES

Alvarez-Ponce D. & J.O. McInerney (2011) 'The human genome retains relics of its prokaryotic ancestry: human genes of archaebacterial and eubacterial origin exhibit remarkable differences', Genome biology and evolution 3: 782–90.

Alvarez-Ponce D., P. Lopez, E. Bapteste & J.O. McInerney (2013) 'Gene similarity networks provide new tools for understanding eukaryote origins and evolution.', Proceedings of the National Academy of Sciences of the United States of America 110(17): E1594–603.

Bapteste, E. et al. (2008) 'Alternative methods for concatenation of core genes indicate a lack of resolution in deep nodes of the prokaryotic phylogeny.', Mol Biol Evol 25(1): 83–91.

Bapteste, E. et al. (2012) 'Evolutionary analyses of non-geological bonds produced by introgressive descent', Proceedings of the National Academy of Sciences of the United States of America 109(45): 18266–72.

Bapteste, E., F. Bouchard & R.M. Burian (2012) 'Philosophy and Evolution : Minding the Gap Between Evolutionary Patterns and Tree-Like Patterns', in M. Anisimova (ed), Evolutionary Genomics: Statistical and Computational Methods, Vol.2 (New York, Humana Press: Springer): 81–112

Bapteste, E., M. O'Malley & R. Beiko (2009) Prokaryotic evolution and the tree of life are two different things, Biol Direct. 4:34.

Beauregard-Racine, J. et al. (2011) 'Of woods and webs: possible alternatives to the tree of life for studying genomic fluidity in E. coli', Biol. Direct 6(1): 39.

Bittner, L. et al. (2010) 'Some considerations for analyzing biodiversity using integrative metagenomics and gene networks', Biol Direct 5:47.

Burian, R.M. (2011) 'Experimentation, Exploratory', in W. Dubitzky, O. Wolkenhauer, K.-H. Cho & H. Yokota, Encyclopedia of Systems Biology (in press).

Dagan, T. & W. Martin (2009) 'Getting a better picture of microbial evolution en route to a network of genomes', Philos Trans R Soc Lond B Biol Sci 364(1527): 2187–96.

Dagan, T., W. Martin (2006) 'The tree of one percent', Genome Biol 7(10): 118.

Dagan, T., Y. Artzy-Randrup & W. Martin (2008) 'Modular networks and cumulative impact of lateral transfer in prokaryote genome evolution', Proceedings of the National Academy of Sciences of the United States of America 105(29): 10039–44.

Downey, S.S., B. Hallmark, M.P. Cox, P. Norquest & S. Lansing (2008) 'Computational feature-sensitive reconstruction of language relationships: Developing the ALINE distance for comparative historical linguistic reconstruction', Journal of Quantitative Linguistics 15(4): 340–69.

Fondi, M. & R. Fani (2010) 'The horizontal flow of the plasmid resistome: clues from inter-generic similarity networks', Environ Microbiol 12(12): 3228–42.

Franklin-Hall, L. (2005) 'Exploratory Experiments', Philosophy of Science 72: 888–99.

Gallagher B. & T. Eliassi-Rad (2008) 'Leveraging Label-Independent Features for Classification in Sparsely Labeled Networks: An Empirical Study.', SNA-KDD '08.

Ghigo, J.M. (2001) 'Natural conjugative plasmids induce bacterial biofilm development', Nature 412(6845): 442–5.

Halary, S., J.W. Leigh, B. Cheaib, P. Lopez & E. Bapteste (2010) 'Network analyses structure genetic diversity in independent genetic worlds', Proceedings of the National Academy of Sciences of the United States of America 107(1): 127–31.

Hall-Stoodley, L., J.W. Costerton & P. Stoodley (2004) 'Bacterial biofilms : from the natural environment to infectious diseases', Nat Rev Microbiol 2(2): 95–108.

Hóu, J. (2004) Xiàndài Hànyu fangyán yinkù [Phonological database of Chinese dialects] (Shanghai: Shànghai Jiàoyu).

Huang, J. & J.P. Gogarten (2007) 'Did an ancient chlamydial endosymbiosis facilitate the establishment of primary plastids ?' Genome Biol 8(6): R99.

Jachiet, P.A., R. Pogorelcnik, A. Berry, P. Lopez & E. Bapteste (2013) 'MosaicFinder: identification of fused gene families in sequence similarity networks', Bioinformatics.

Jones, B.V. (2010) 'The human gut mobile metagenome: a metazoan perspective', Gut Microbes 1(6): 415–31.

Kloesges, T., O. Popa, W. Martin & T. Dagan (2011) 'Networks of gene sharing among 329 proteobacterial genomes reveal differences in lateral gene transfer frequency at different phylogenetic depths', Mol Biol Evol 28(2): 1057–74.

Koschützki, D. (2008) 'Network Centralities', in B.H. Junker & F. Schreiber, Analysis of Biological Networks (Hoboken, NJ: John Wiley & Sons, Inc.): 65–84.

Lass, R. (1997) Historical linguistics and language change (Cambridge: Cambridge University Press).

Leigh, J.W., F.J. Lapointe, P. Lopez & E. Bapteste (2011) 'Evaluating phylogenetic congruence in the post-genomic era', Genomic biology and evolution 3: 571–87.

Leskovec, J., K.J. Lang, A. Dasgupta & M.W. Mahoney (2008) 'Statistical properties of community structure in large social and information networks.', Proc. 17-th International WWW: 695–704.

Levenshtein, V.I. (1966) 'Binary codes capable of correcting deletions, insertions, and reversals', Soviet Physics Doklady 10(8): 707–10.

Lima-Mendez, G., J. Van Helden, A. Toussaint & R. Leplae (2008) 'Reticulate representation of evolutionary and functional relationships between phage genomes', Mol Biol Evol 25(4):762–77.

List, J.-M. (2012) 'LexStat. Automatic Detection of Cognates in Multilingual Wordlists.', Joint Workshop of LINGVIS & UNCLH: 117–25.

List, J.-M. (2012) 'SCA. Phonetic alignment based on sound classes', in M. Slavkovik & D. Lassiter (eds), New directions in logic, language and computation (Berlin/Heidelberg: Springer): 32–51.

Lozupone, C.A. et al. (2008) 'The convergence of carbohydrate active gene repertoires in human gut microbes', Proceedings of the National Academy of Sciences of the United States of America 105(39): 15076–81.

Marin, B., E.C. Nowack & M. Melkonian (2005) 'A plastid in the making: evidence for a second primary endosymbiosis', Protist 156(4): 425–32.

Martha, V.S. et al. (2011) 'Constructing a robust protein-protein interaction network by integrating multiple public databases', BMC Bioinformatics 12 Suppl. 10:S7.

Martin, W. et al. (2007) 'The evolution of eurakyotes', Science 316(5824): 542–3.

Moustafa, A. et al. (2009) 'Genomic footprints of a cryptic plastid endosymbiosis in diatoms', Science 324(5935): 1724–6.

Nelson-Sathi, S. et al. (2011) 'Networks uncover hidden lexical borrowing in Indo-European language evolution', Proceedings. Biological sciences / The Royal Society 278(1): 1794–803.

O'Malley, M.A. & E.V. Koonin (2011) 'How stands the Tree of Life a century and a half after The Origin?' Biol Direct 6: 32.

Periasamy, S. & P.E. Kolenbrander (2009) 'Aggregatibacter actinomycetemcomitans builds mutualistic biofilm communities with Fusobacterium nucleatum and Veillonella species in salvia', Infect Immun 77(9): 3542–51.

Qu, A. et al. (2008) 'Comparative metagenomics reveals host sprecific metavirulomes and horizontal gene transfer elements in the chicken cecum microbiome', PloS One 3(8): e2945.

Skippington, E. & M.A. Ragan (2011) 'Lateral genetic transfer and the construction of genetic exchange communities', FEMS Microbiol Rev 35(5): 707–35.

Smoot, M.E., K. Ono, J. Ruscheinski, P.L. Wang & T. Ideker (2011) 'Cytoscape 2.8: new features for data integration and network visualization', Bioinformatics 27(3): 431–2.

Swadesh, M. (1952) 'Lexicostatic dating of prehistoric ethnic contacts', Proceedings American Philosophical Society 96: 452–63.

Swadesh, M. (1955) 'Towards greater accuracy in lexicostatistic dating', International Journal of American Linguistics 21: 121–37.

Turchin P., I. Peiros & M. Gell-Mann (2010) 'Analyzing genetic connections between languages by matching consonant classes', Journal of Language Relationship 3: 117–26.

Vinayagam A. et al. (2011) 'A directed protein interaction network for investigating intracellular signal transduction.', SciSignal 4(189): rs8.

Wang, T.Y., F. He, Q.W. Hu & Z. Zhang (2011) 'A predicted protein-protein interaction network of the filamentous fungus Neurospora crasse', Mol Biosyst 7(7): 2278–85.

Wintermute, E.H. & P.A. Silver (2010) 'Dynamics in the mixed microbial concourse', Genes Dev 24(23): 2603–14.

Wu, D. et al. (2011) 'Stalking the fourth domain in metagenomic data: searching for, discovering, and interpreting novel, deep branches in marker gene phylogenetic trees', PloS One 6(3): e18011.

Zhaxybayeva. O. & F. Doolittle (2010) 'Metagenomics and the Units of Biological Organization', BioScience 60(2): 102–12.

ACKNOWLEDGEMENTS

We thank Thorsten Halling for initiating collaboration between the three of us, and Bill Martin and Tal Dagan for their pioneering work in linguistics using language networks.

AUTHORS

Andersen, Hanne
Department of Physics and Astronomy – Science Studies, Aarhus University, Danmark
E-Mail: hanne.andersen@ivs.au.dk

Bapteste, Eric
Unité de recherche Systématique, Adaptation, Évolution (UMR 7138), Université Pierre et Marie Curie Paris, France
E-Mail: epbapteste@gmail.com

Dagan, Tal
Institute of Microbiology, Christian-Albrechts-University Kiel, Germany
Institute of Genomic Microbiology, Heinrich-Heine University Düsseldorf, Germany (until 2012)
E-Mail: tdagan@ifam.uni-kiel.de

Fangerau, Heiner
Institute of the History, Philosophy and Ethics of Medicine, University of Ulm, Germany
E-Mail: heiner.fangerau@uni-ulm.de

Geisler, Hans
Institute of Romance Languages and Literature, Heinrich-Heine University Düsseldorf, Germany
E-Mail: geisler@phil.uni-duesseldorf.de

Halling, Thorsten
Institute of the History, Philosophy and Ethics of Medicine, University of Ulm, Germany
E-Mail: thorsten.halling@uni-ulm.de

Hoquet, Thierry
Institut de Recherches Philosophiques de Lyon (IRPHIL)
E-Mail: thierry.hoquet@univ-lyon3.fr

Kressing, Frank
Institute of the History, Philosophy and Ethics of Medicine, University of Ulm, Germany
E-Mail: frank.kressing@uni-ulm.de

Krischel, Matthis
Institute of the History, Theory and Ethics of Medicine, RWTH Aachen University,
Germany
E-Mail: mkrischel@ukaachen.de

List, Johann-Mattis
Forschungszentrum Deutscher Sprachatlas of Philipps-University Marburg, Ger-
many
Institute of Romance Languages and Literature, Heinrich-Heine University, Düs-
seldorf, Germany (until 2012)
E-Mail: mattis.list@uni-marburg.de

Lopez, Philippe
Unité de recherche Systématique, Adaptation, Évolution (UMR 7138), Université
Pierre et Marie Curie Paris, France
E-Mail: philippe.lopez@upmc.fr

Martin, William F.
Institute of Molecular Evolution, Heinrich-Heine University Düsseldorf, Germany
E-Mail: bill@hhu.de

Nelson-Sathi, Shijulal
Institute of Molecular Evolution, Heinrich-Heine University Düsseldorf, Germany
E-Mail: shijulalns001@gmail.com

Nerbonne, John
Humanities Computing University of Groningen, Netherlands
E-Mail: j.nerbonne@rug.nl

Popa, Ovidiu
Institute of Genomic Microbiology, Heinrich-Heine University Düsseldorf, Ger-
many
E-Mail: ovidiu.popa@uni-duesseldorf.de

Prokić, Jelena
Research Unit "Quantitative Language Comparison", LMU Munich, Germany
E-Mail: j.prokic@lmu.de

Starostin, George
Institute for Oriental and Classical Studies, Russian State University for the Hu-
manities, Moscow
E-Mail: gstarst@rinet.ru

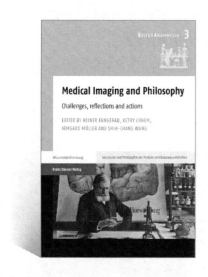

Franz Steiner Verlag

Heiner Fangerau / Rethy K. Chhem /
Irmgard Müller / Shih-Chang Wang (ed.)
Medical Imaging and Philosophy

2012.
190 Seiten mit 31 Abbildungen.
Kart.
ISBN 978-3-515-10046-5

Heiner Fangerau / Rethy K. Chhem /
Irmgard Müller / Shih-Chang Wang (ed.)

Medical Imaging and Philosophy

Challenges, reflections and actions

Kulturanamnesen – Band 3

Medical Imaging plays a prominent role in contemporary
medical research and practice. At the same time imaging in
its broadest sense, including illustration, diagramming,
model-making, photography and other forms of image
rendering, has a long tradition in medicine. Imaging the
human body has different aspects not only related to techniques or current interpretations of visual representations
through medical imaging technologies. The way the human
body was and is displayed in medicine also reflects a range
of cultural, historical, artistic and scientific concerns.
This book summarizes the results of an interdisciplinary
conference on medical imaging. It offers stimulating
papers giving perspectives on Medical Imaging from
medicine, philosophy, history and arts.

Aus dem Inhalt

Franz Steiner Verlag
Birkenwaldstr. 44 · D – 70191 Stuttgart
Telefon: 0711 / 2582 – 0 · Fax: 0711 / 2582 – 390
E-Mail: service@steiner-verlag.de
Internet: www.steiner-verlag.de

Heiner Fangerau /
Igor J. Polianski (Hg.)
Medizin im Spiegel ihrer Geschichte,
Theorie und Ethik

2012.
264 Seiten mit 35 Abbildungen.
Kart.
ISBN 978-3-515-10227-8

Heiner Fangerau / Igor J. Polianski (Hg.)

Medizin im Spiegel ihrer Geschichte, Theorie und Ethik

Schlüsselthemen für ein junges Querschnittsfach

Kulturanamnesen – Band 4

Die Novelle der Ärztlichen Approbationsordnung von 2002 sah für das Medizinstudium die curriculare Etablierung der Medizinethik zusammen mit der Medizingeschichte und der Medizintheorie in einem Querschnittsfach vor. Die Schaffung dieser Trinität war jedoch umstritten. Blickt man heute auf die Debatte zurück, so ist festzustellen, dass sich mittlerweile mit dem „Dreigestirn" von Geschichte, Theorie und Ethik der Medizin (GTE) ein Fachzuschnitt und eine Fachkultur formiert hat, wie es sie zu Beginn der Debatte so noch nicht gegeben hatte. Dieser Band schließt vor diesem Hintergrund an die bisherigen Reflexionen an, bietet eine aktuelle Standortbestimmung von GTE und erhellt die Perspektiven dieses Querschnittsfaches.

Aus dem Inhalt

C. WIESEMANN: Die Beziehung der Medizinethik zur Medizingeschichte und Medizintheorie | I. J. POLIANSKI / H. FANGERAU: Legitimationsdruck und Theoretisierungszwang | V. ROELCKE: Medizin im Nationalsozialismus – radikale Manifestation latenter Potentiale moderner Gesellschaften? | W. BRUCHHAUSEN: Globalisierungsgeschichte der Medizin am Beispiel der deutschen Entwicklungshilfe | I. MÜLLER: Macht und Evidenz der Bilder in J. Zahns „Oculus artificialis" | H. SCHOTT: Die „natürliche Magie" und ihre Bedeutung für die Medizingeschichte | G. BADURA-LOTTER: Sexuell übertragbare Krankheiten. Überlegungen zu Metaphorik und Ethik | E. BRINKSCHULTE: Zur „Gehörlosenproblematik" seit dem 18. Jahrhundert | S. KESSLER: Diskursive Konstruktion von sozialer Ungleichheit und Krankheit in Deutschland | D. L. FROMMELD: Der Body-Mass-Index (BMI) als biopolitisches Instrument | J. VÖGELE: Säuglingsfürsorge, Säuglingsernährung und die Entwicklung der Säuglingssterblichkeit in Deutschland während des 20. Jahrhunderts | u.a.

Franz Steiner Verlag
Birkenwaldstr. 44 · D – 70191 Stuttgart
Telefon: 0711 / 2582 – 0 · Fax: 0711 / 2582 – 390
E-Mail: service@steiner-verlag.de
Internet: www.steiner-verlag.de